AA003003

Computational Molecular Science and Engineering Forum 2023

Held at the 2023 AIChE Annual Meeting

Orlando, Florida, USA
5-10 November 2023

ISBN: 978-1-7138-9301-1

Printed from e-media with permission by:

Curran Associates, Inc.
57 Morehouse Lane
Red Hook, NY 12571

proceedings.com

Some format issues inherent in the e-media version may also appear in this print version.

Printed with permission by Curran Associates, Inc. (2024)

For permission requests, please contact AIChE
at the address below.

AIChE
120 Wall Street, FL 23
New York, NY 10005-4020

Phone: (800) 242-4363
Fax: (203) 775-5177

www.aiche.org

Additional copies of this publication are available from:

Curran Associates, Inc.
57 Morehouse Lane
Red Hook, NY 12571 USA
Phone: 845-758-0400
Fax: 845-758-2633
Email: curran@proceedings.com
Web: www.proceedings.com

TABLE OF CONTENTS

AUTOMATED MOLECULAR AND MATERIALS DISCOVERY: INTEGRATING MACHINE LEARNING, SIMULATION, AND EXPERIMENT

PLENARY SESSION: COMPUTATIONAL MOLECULAR SCIENCE AND ENGINEERING FORUM (INVITED TALKS)

APPLICATIONS OF MOLECULAR MODELING TO STUDY INTERFACIAL PHENOMENA

RECENT ADVANCES IN MOLECULAR SIMULATION METHODS

PRACTICAL APPLICATIONS OF COMPUTATIONAL CHEMISTRY AND MOLECULAR SIMULATION

SOFTWARE ENGINEERING IN AND FOR THE MOLECULAR SCIENCES

SPOTLIGHTS IN THERMODYNAMICS AND COMPUTATIONAL MOLECULAR SCIENCE (INVITED TALKS)

MACHINE LEARNING FOR SOFT AND HARD MATERIALS

RECENT ADVANCES IN MULTISCALE METHODOLOGIES

POSTER SESSION: COMPUTATIONAL MOLECULAR SCIENCE AND ENGINEERING FORUM

RECENT ADVANCES IN MOLECULAR SIMULATION METHODS II

APPLICATIONS OF MOLECULAR MODELING TO STUDY INTERFACIAL PHENOMENA II

MACHINE LEARNING FOR SOFT AND HARD MATERIALS II

AUTOMATED MOLECULAR AND MATERIALS DISCOVERY: INTEGRATING MACHINE LEARNING, SIMULATION, AND EXPERIMENT II

361a Efficient Data-Driven Discovery of Novel Innate Immunomodulators Using Machine Learning-Guided High Throughput Screening

Yifeng Tang, Jeremiah Kim, Carman Ka Man Ip, Azadeh Bahmani, Qing Chen, Matthew Rosenberger, Aaron Esser-Kahn, Andrew Ferguson

The success of prophylactic vaccines and immunotherapies relies on the innate immune response, which can be modulated by small molecules to inhibit unfavorable systemic inflammation for prophylactic vaccines and improve immune stimulation for immunotherapies to mitigate suppression from tumor micro-environment. Traditional screening methods for discovering small molecule with specific pharmaceutical usages are time-consuming and resource-intensive, and often fail to identify novel compounds with high efficacy. In addition, although human intuition and experience present valuable heuristics to guide this search, the relative infancy of immunomodulator discovery efforts, absence of mechanistic understanding, and vast size of molecular space can make these heuristics limiting and subject to human preconceptions, bias, and potential blind spots.

We have developed a machine learning-enabled active learning pipeline to guide in vitro experimental screening and discovery of small molecule immunomodulators that alter the signaling activity of innate immune responses stimulated by traditional pattern recognition receptor agonists. Molecules were tested by in vitro high throughput screening (HTS) where we measured modulation over the activation of two major effector of human immunity, the nuclear factor κ-light-chain-enhancer of activated B-cells (NF-κB) pathway and the interferon regulatory factors (IRF) pathway. These data were used to train data-driven predictive models linking molecular structure to their immunomodulatory activity using deep representational learning, Gaussian process regression (GPR), and Bayesian optimization (BO). The deep representational learning model was trained to convert molecular structure to continuous embeddings with 97% reversible fidelity. The GPR and BO converged after four rounds of discovery loop between computation and experimentation with stabilized prediction and minimized error.

By interleaving successive rounds of model training and in vitro HTS, we performed an active learning-guided traversal of a 139,998-molecule library using a fraction of the time and material costs associated with exhaustive screening. After experimentally evaluating only around 2% of the library, we discovered molecules with unprecedented immunomodulatory capacity, including those capable of suppressing NF-κB activity by up to 15-fold, elevating NF-κB activity by up to 5-fold, and elevating IRF activity by up to 3-fold. A subset of our top-performing candidates, namely 17 compounds, was tested to validate their immunomodulatory effect by measuring cytokine release profiles. One of these molecules demonstrated a 40-fold enhancement in IFN-β production. In addition, we rationalized the correlation between the chemical structure and immunomodulatory capacity by using a linear regression model and interpretable molecular features and discovered design rules for immunomodulators.

Our machine learning-enabled screening approach presents an efficient immunomodulator discovery pipeline that has furnished a library of novel small molecules with unprecedented capacity to enhance or suppress innate immune signaling pathways. This has the potential to improve prophylactic vaccination by minimizing side effects and addressing vaccination hesitancy, and to enhance the potency of immunotherapies. This collection of new small molecule immunomodulators may progress to subsequent screenings, in vivo studies, clinical trials, and eventually become a pharmaceutical product, if successful. Furthermore, this machine learning-based screening strategy can be employed in drug discovery and development, especially in pipelines lacking well-defined mechanistic insights and requiring expensive experimental assessments.

361b Automated, Hybrid Model-Driven pH Adjustment of Viscous Liquid Formulations

Aniket Chitre, Jayce Cheng, Kedar Hippalgaonkar, Alexei A. Lapkin

Liquid-formulated consumer products are adjusted to a target pH physiologically compatible with our bodies. For these complex, viscous mixtures, the titration process is carried out manually in a painstakingly slow trial-and-error process by experienced industrial scientists. Herein, we accelerate the pH adjustment process by i) building a new robotic platform capable of titrating these challenging viscous formulations, ii) developing further the active machine learning (ML)-driven pH adjustment strategy presented by Pomberger et al., 2023. The platform's novel design with a 3d-printed overhead stirrer and pH probe in a single position facilitates the titration of viscous samples. Additionally, a washing station involving a soak & jetting down in isopropyl alcohol enables full automation as the robot moves between the closed-loop pH adjustment of different samples. Consumer products are typically titrated with citric acid, as this weak, polyprotic acid can confer the final product a strong buffering effect to remain at the target pH, resulting in a longer shelf-life. However, the prior literature has only developed an ML strategy compatible with strong-strong acid/base titrant pairs. Here a weak-strong acid/base pair is considered, where the deprotonation state of the weak acid depends on the pH, and there is a possibility of forming a complex buffer solution if the target pH is overshot. Accordingly, a data-driven approach (Gaussian Process regression) is combined with physical chemistry to present a hybrid model capable of efficiently guiding the robotic platform to the target pH. Finally, scale-up studies are underway, with the aim that this technology is adopted into industrial high-throughput formulations workflows.

361c Autonomous Construction of Nanomaterial Phase Maps Using High-Throughput X-Ray Scattering

Kiran Vaddi, Karen Li, Sage Scheiwiller, Lilo Pozzo

Automation in many of the experimental pipelines both at laboratory and central facilities has shifted the bottleneck to autonomously-driven data analysis and decision-making. Exponential growth in tools available for data-driven modeling resulted in the advent of self-driven laboratories (SDL) that aims to automate and accelerate the entire workflow starting from synthesis to characterization and device integration for emerging technologies and energy needs. Platforms based on solution-processible materials (polymers, colloids, and nanoparticles) are amenable to automation both at the synthesis and characterization levels. Techniques such as scattering and spectroscopy provide faster high-throughput alternatives to capture a signal of the underlying structure allowing us to construct composition-structure phase maps. Although obtaining a structure phase map from high throughput structural characterization data is a decade-old problem, many of the state-of-the-art approaches are not generalizable beyond their initial application in X-ray diffraction. In particular, small-angle scattering (SAS) used to probe the structure of nanoscale material results in a continuous signal where the key information is not just in peak positions but also the overall 'shape' of the curve. Thus, self-driven laboratories for soft materials should be equipped with tools that can process the shape of SAS curves into useful information for decision-making and optimization. In this talk, we present a novel approach that uses Riemannian metric geometry to develop a shape-based measure for analyzing SAS curves. We use this novel analysis tool in an SDL to build a phase map of the block copolymer system in an iterative closed-loop fashion. Our approach allows the decision-making to be built on top of multiple Riemannian metric fields each corresponding to an order parameter of the phase map. Using a case study of a novel polymer blend, we show that the learned order parameters are continuous and thus have an interpretation in terms of multi-phase transitions.

361d Pareto Optimization to Accelerate Multi-Property Virtual Screening

Jenna Fromer, David Graff, Connor Coley

Molecular discovery is a constrained multi-objective optimization problem that aims to identify one or more molecules that balance multiple, often competing, properties. One approach to molecular design is to evaluate the properties of all molecules in a virtual library ("virtual screening"), and methods based on active learning and Bayesian optimization have previously been proposed to accelerate virtual screening [1, 2]. However, these methods typically consider single-objective optimization only or, in the case of multi-objective optimization, combine properties of interest into a scalarized objective [3]. A promising alternative to scalarization is Pareto optimization, which can elucidate the trade-offs between objectives and does not rely on assumptions about the relative importance of objectives. In the context of molecular discovery, Pareto optimization identifies the molecules that optimally balance multiple desired properties, revealing the best that is possible in the objective space.

We combine Pareto optimization and Bayesian optimization methods to efficiently search a library of millions of molecules and identify those that optimally balance multiple objectives. Using graph neural networks as the surrogate model architecture, we apply this methodology to identify (1) dual inhibitors which minimize docking scores to two targets and (2) selective small molecule inhibitors which minimize docking scores to an on-target and maximize those to an off-target. We compare optimization performance between Pareto and scalarization acquisition functions and find that multi-objective acquisition functions outperform scalarization. This improvement is greater for the selective inhibitor task due to stronger competition between the objectives. Finally, we find that encouraging molecular and functional diversity during acquisition does not improve the hypervolume of acquired molecules but does increase the diversity of the acquired set.

[1] Gentile, F.; Agrawal, V.; Hsing, M.; Ton, A.-T.; Ban, F.; Norinder, U.; Gleave, M. E.; Cherkasov, A. Deep Docking: A Deep Learning Platform for Augmentation of Structure Based Drug Discovery. *ACS Cent. Sci.* **2020,** *6* (6), 939–949.

[2] Graff, D. E.; Shakhnovich, E. I.; Coley, C. W. Accelerating High-Throughput Virtual Screening Through Molecular Pool-Based Active Learning. *Chem. Sci.* **2021,** *12* (22), 7866–7881.

[3] Mehta, S.; Goel, M.; Priyakumar, U. D. MO-MEMES: A Method for Accelerating Virtual Screening Using Multi-Objective Bayesian Optimization. *Frontiers in Medicine* **2022,** *9,* 916481.

361e Quantumpioneer: Self-Evolving Machine for High-Throughput Automated Potential Energy Surface Exploration and Closed-Loop Chemical Reactivity Discovery

Haoyang Wu, Hao-Wei Pang, Xiaorui Dong, Jackson W. Burns, Kevin Spiekermann, Angiras Menon, Jonathan Zheng, Sayandeep Biswas, Shih-Cheng Li, William Green

Acquiring accurate 3D geometries of reactive chemical species and transition states (TS) along reaction pathways over the potential energy surface is a critical prerequisite for computing high-quality thermodynamic and kinetic parameters via quantum chemistry and statistical mechanics. These calculated thermo-kinetic parameters are essential for creating impactful predictive models that can assist pharmaceutical design, high-performance material development, and renewable energy research. However, the search for transition states has historically been a trial-and-error process that demands considerable manual input. Although automated TS search has a rich history of algorithmic development, existing methods like single-ended and double-ended approaches are often computationally demanding, have high failure rates, and are highly localized to specific reaction systems, making it challenging to learn from previous examples and enhance performance. Therefore, efficiently automating the TS search is essential for enabling high-throughput chemical reactivity discovery and streamlining the generation of extensive theoretical reaction datasets that are valuable for developing machine-learning-based chemistry surrogate models. In this presentation, we introduce QuantumPioneer, a self-evolving machine capable of automatically generating validated 3D electronic structures for a given reaction using atom-mapped SMILES as input. QuantumPioneer can enhance its performance over time by learning from past experiences and can be seamlessly integrated into an active learning loop for closed-loop chemical discovery. Moreover, QuantumPioneer can be tailored to specific chemical-exploration tasks and generate extensive theoretical reaction datasets that lay the foundation for future studies with wide-ranging applications. To showcase its capabilities, we employed QuantumPioneer to generate TS at ωB97XD/def2svp level of theory for over 100,000 hydrogen atom abstraction (HAbs) reactions between peroxyl radicals and a diverse set of organic molecules chosen based on their functional groups. This dataset represents one of the largest DFT TS collections for radical-based reactions to date, laying the foundation for future studies with wide-ranging applications. We further refined the dataset with COSMO-RS and DLPNO-CCSD(T1)-F12D single-point calculations and trained D-MPNN models to predict HAbs reaction barriers. We will demonstrate how the machine learning model for HAbs barrier height prediction can potentially be applied to conduct high-throughput virtual screening of oxidative stability of organic molecules and aid better design of high-value chemicals such as active pharmaceutical ingredients and carbon capture amines.

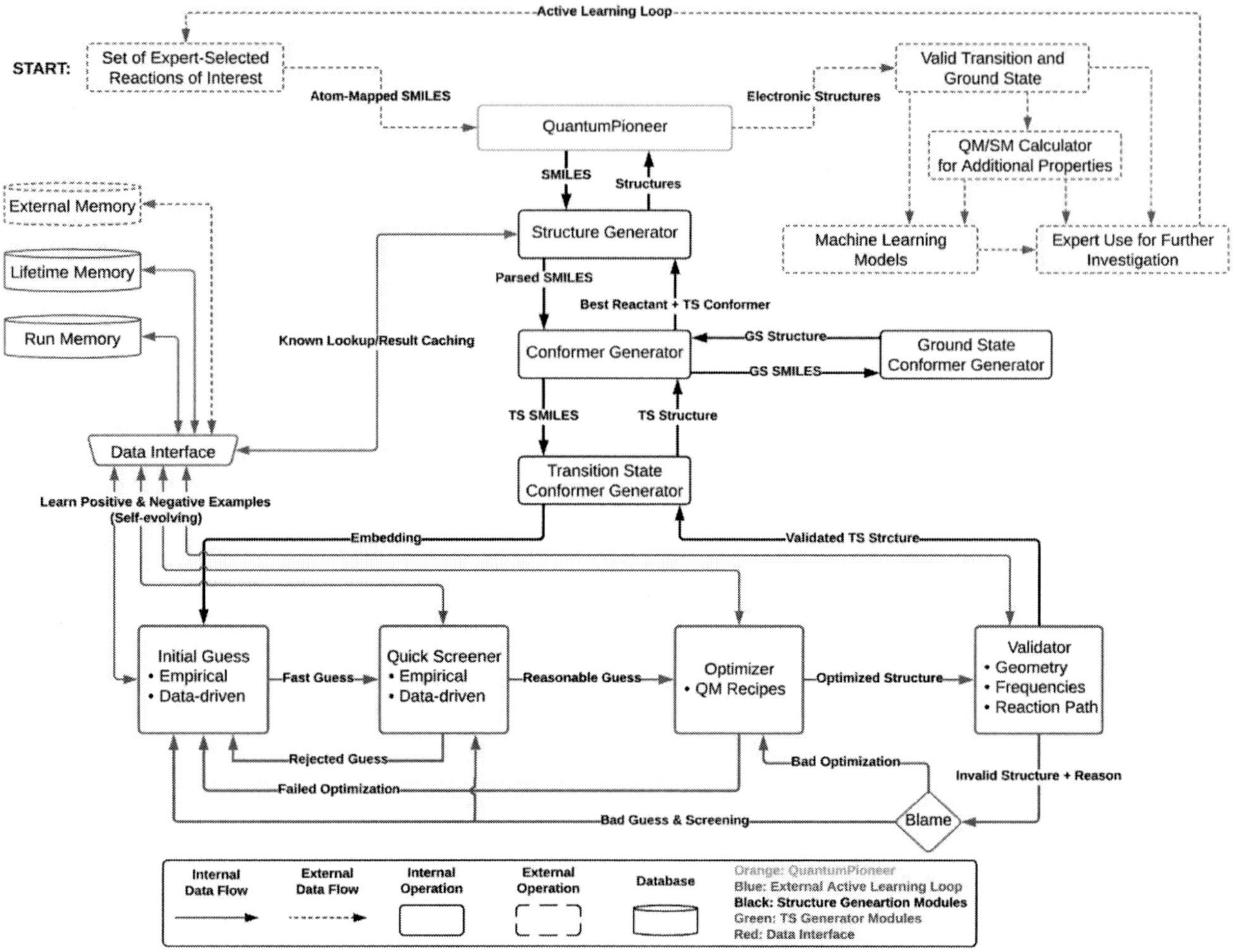
Active Learning Loop
START:
Set of Expert-Selected Reactions of Interest
Atom-Mapped SMILES
QuantumPioneer
Electronic Structures
Valid Transition and Ground State
QM/SM Calculator for Additional Properties
Machine Learning Models
Expert Use for Further Investigation
SMILES
Structures
External Memory
Lifetime Memory
Run Memory
Structure Generator
Parsed SMILES
Best Reactant + TS Conformer
Known Lookup/Result Caching
Conformer Generator
GS Structure
GS SMILES
Ground State Conformer Generator
TS SMILES
TS Structure
Data Interface
Transition State Conformer Generator
Learn Positive & Negative Examples (Self-evolving)
Embedding
Validated TS Strcture
Initial Guess
• Empirical
• Data-driven
Fast Guess
Quick Screener
• Empirical
• Data-driven
Reasonable Guess
Optimizer
• QM Recipes
Optimized Structure
Validator
• Geometry
• Frequencies
• Reaction Path
Rejected Guess
Bad Optimization
Invalid Structure + Reason
Failed Optimization
Bad Guess & Screening
Blame
Internal Data Flow
External Data Flow
Internal Operation
External Operation
Database
Orange: QuantumPioneer
Blue: External Active Learning Loop
Black: Structure Geneartion Modules
Green: TS Generator Modules
Red: Data Interface

361f Design of Peptide-Based Drugs for Treating *Clostridioides Difficile* Infection

Sudeep Sarma, Carly Catella, Deniz Durmusoglu, Nathan Crook, Scott Magness, Stefano Menegatti, Carol Hall

Clostridioides difficile infection is the leading cause of diarrhea and colitis (inflammation of the colon) in North America and Europe. The fraction of the population infected by the disease is increasing as new strains associated with significant morbidity and mortality have appeared. There is a growing concern about the failure of the first line of treatment for *C. diff.* infection (metronidazole and vancomycin) and thus, current attention is focused on the need for alternative treatment options like biologic drugs which are safer and more efficacious. A viable and cost-effective strategy is to develop targeted peptide-based therapeutics that prevent and treat *C. diff.* infections by deactivating the pathogen while leaving the resident gut microbiota unharmed.

During infection, the *C. diff.* pathogen produces two large virulent toxins (toxins A and B) that share 71% sequence homology. The glucosyltransferase domain (GTD) of these toxins acts by binding uridine diphosphate (UDP)-glucose, hydrolyzing it into glucose and UDP, and attaching the glucose monomer to the human Rho family of GTPases. Glycosylation of the GTPases by toxin A and toxin B GTDs leads to disruption of the cytoskeleton, apoptosis, and death of the colon epithelial cells.

The objective of this project is to computationally design peptide inhibitors that bind with high affinity and specificity to *C. diff.* toxin A GTD and toxin B GTD to inhibit its activity. We use an automated ***Pep***tide ***B***inding ***D***esign (PepBD) algorithm developed in our lab to design peptide sequences that bind to *C. diff.* Toxin A. Our peptide search algorithm employs Monte Carlo methods, self-consistent mean-field theory, and the concerted rotation (CONROT) technique to search for peptides in sequence and conformation space that bind to a target protein. The starting input structure for the algorithm requires a reference ligand, which is usually an experimentally determined peptide sequence that has a good binding affinity to the target protein.

We identified an 8-mer peptide, SA1, that binds to the *Clostridioides difficile* toxin A GTD. The efficacy of peptide SA1 was tested using a trans-epithelial electrical resistance (TEER) assay on monolayers of the human gut epithelial culture model. Peptide SA1 blocks TcdA toxicity in jejunum (small intestine) cells and in colon epithelial cells and exhibits a binding affinity in the nanomolar range. Next, we designed peptides that bind to toxin B GTD. Preliminary experimental data suggest that two of our *in-silico* identified peptides, SB2 and SB6, are inhibiting glycosylation of toxin B GTD with an IC50 in the micromolar range. Collectively, our results demonstrate the potential of our computational peptide design protocol to identify peptide-based inhibitors that can treat *Clostridoides difficile* infection and other infectious pathogens in the future.

361g From Platform to Knowledge Graph: Distributed Self-Driving Laboratories

Jiaru Bai, Sebastian Mosbach, Connor Taylor, Dogancan Karan, Kok Foong Lee, Simon Rihm, Jethro Akroyd, Alexei A. Lapkin, Markus Kraft

The ability to integrate resources and share knowledge across organisations enables scientists to expedite the scientific discovery process, which is especially crucial in addressing emerging global challenges that require global solutions [1, 2]. In this work, we develop an architecture to enable distributed self-driving laboratories as part of The World Avatar project, an all-encompassing digital twin based on dynamic knowledge graph. Our approach utilises ontologies to capture the data and material flows involved in a design-make-test-analyse cycle, and employs autonomous agents as executable knowledge components to carry out the experimentation workflow. All data provenance is recorded following FAIR principles, ensuring its accessibility and interoperability. We demonstrate the practical application of our framework by linking two robotic setups in Cambridge and Singapore to achieve a collaborative closed-loop optimisation for an aldol condensation reaction in real time. The knowledge graph evolves on its own while progressing towards the research goals set by the scientist. The two robots effectively produced a Pareto front for the cost-yield optimisation problem over the course of two days of operation. This proof-of-concept demonstration highlights the potential of the framework to establish a globally collaborative research network and further advance scientific frontiers.

[1] Seifrid, M. *et al.* Autonomous Chemical Experiments: Challenges and Perspectives on Establishing a Self-Driving Lab. *Acc. Chem. Res.* 55, 2454–2466 (2022).

[2] Bai, J. *et al.* From Platform to Knowledge Graph: Evolution of Laboratory Automation. *JACS Au* 2, 292–309 (2022).

Figure 1: Overall dynamic knowledge graph approach enabling a distributed network of self-driving laboratories.

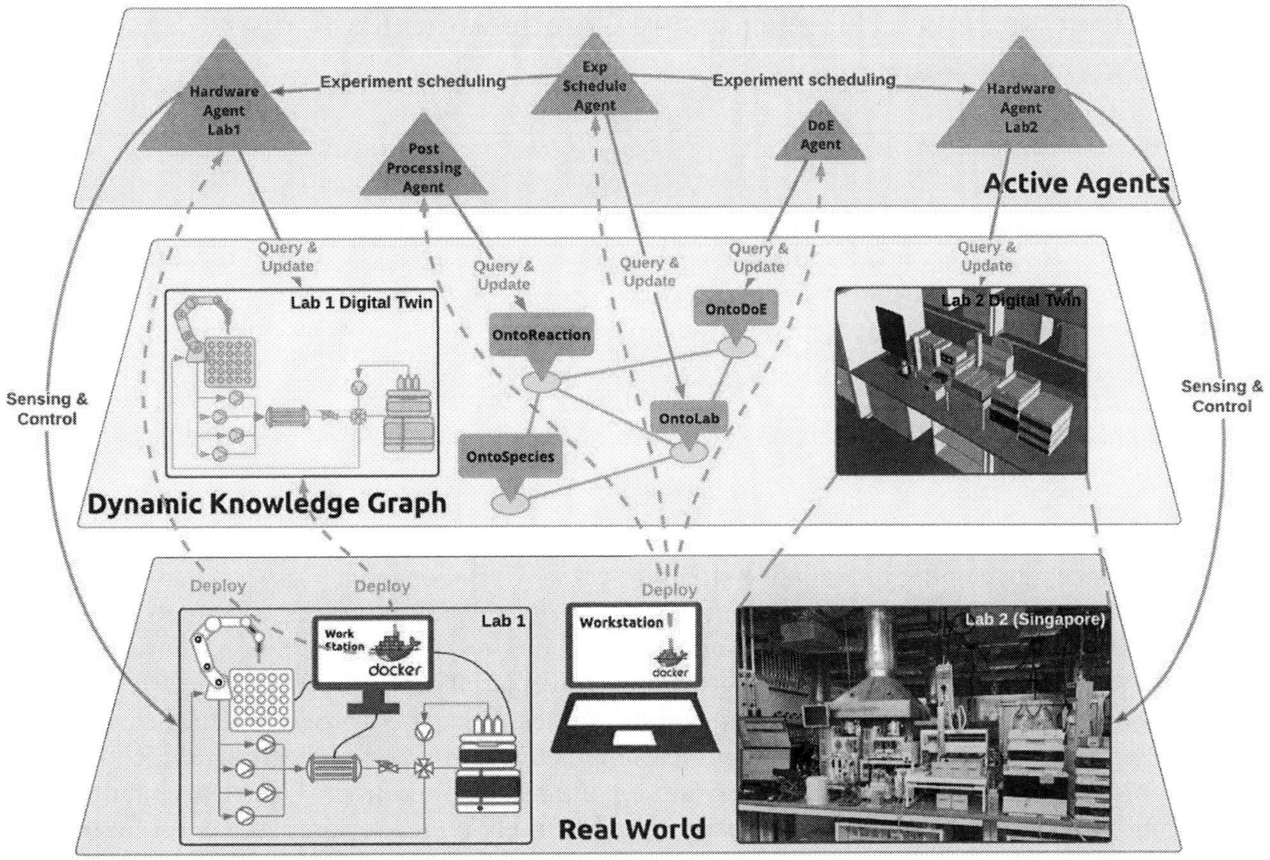

Active Agents
Hardware Agent Lab1
Experiment scheduling
Exp Schedule Agent
Experiment scheduling
Hardware Agent Lab2
Post Processing Agent
DoE Agent
Query & Update
Query & Update
Query & Update
Query & Update
Query & Update
Lab 1 Digital Twin
Lab 2 Digital Twin
OntoReaction
OntoDoE
OntoLab
OntoSpecies
Dynamic Knowledge Graph
Sensing & Control
Sensing & Control
Deploy
Deploy
Deploy
Lab 1
Work Station
docker
Workstation
docker
Lab 2 (Singapore)
Real World

361h Discovering Novel Materials Using Combinatorial Chemistry-Based Reinforcement Learning

Haeyeon Choi, Hyunseung Kim, Dongju Kang, Won Bo Lee, Jonggeol Na

The discovery of new materials with desired properties is important because it can lead to significant advancements in technology and applications. Moreover, the discovery of these new materials tends to have properties that are hard to coexist with or have a very high or low value of specific properties. For example, in the case of transparent electrodes, two properties, transparent and electric, are challenging to exist simultaneously. Here, the problem is that in contrast to conventional materials, few (or any) known samples exhibit these characteristics. Because of this, chemists find it challenging to gain insights or knowledge from the existing materials that might be used to derive the molecular structures of the desired materials.

In recent years, machine learning has provided a new methodology for screening novel materials with high performance, developing quantitative structure-activity relationships (QSARs) and other models, predicting the properties of materials, and finding new materials [1]. However, since most machine learning-based models are probability distribution-based models, such as conditional variational autoencoders (cVAEs) [2] and conditional generative adversarial networks (cGANs) [3], there is a limit to finding materials with extreme characteristic values, since there are very few corresponding molecular samples. In contrast, reinforcement learning is a type of machine learning which learns policy (or order of action) that maximizes reward among selectable actions. Unlike probability distribution-based models, reinforcement learning relies solely on the reward, so it does not matter if a few existing molecules satisfy the target property. Combinatorial chemistry is a methodology that can generate a large number of molecules by combining fragments according to chemical rules. An example of combinatorial chemistry is the Breaking of retrosynthetically interesting chemical substructures (BRICS) [4]. Applying the BRICS rule, molecules can be broken into fragments, or new molecular structures can be generated by combining those fragments.

In this study, we present reinforcement learning to provide combinatorial chemistry a strategy for choosing molecular fragments that directs the molecule that is generated toward the target. With a randomly selected initial fragment, our model selects the next fragment to be combined, considering the chemical rule so that the model can only generate chemically valid molecules. The policy is learned by giving a higher reward if the target properties are satisfied. This learned policy efficiently searches large chemical spaces using combinatorial chemistry. We empirically showed that our model is suitable for discovering novel materials by conducting experiments on several cases, and compared the result with two probability distribution-learning models, conditional recurrent neural network (cRNN) [5] and generative chemical transformer (GCT) [6]. Moreover, after showing that our model performs well by implementing it to a simple problem, we discuss how to improve the performance of the reinforcement learning model and implement the model in real-world problems.

[1] Liu, Y., et al., *Materials discovery and design using machine learning.* Journal of Materiomics, 2017. **3**(3): p. 159-177.

[2] Sohn, K., H. Lee, and X. Yan, *Learning structured output representation using deep conditional generative models.* Advances in neural information processing systems, 2015. **28**.

[3] Mirza, M. and S. Osindero, *Conditional generative adversarial nets.* arXiv preprint arXiv:1411.1784, 2014.

[4] Degen, J., et al., *On the Art of Compiling and Using'Drug-Like'Chemical Fragment Spaces.* ChemMedChem: Chemistry Enabling Drug Discovery, 2008. **3**(10): p. 1503-1507.

[5] Kotsias, P.-C., et al., *Direct steering of de novo molecular generation with descriptor conditional recurrent neural networks.* Nature Machine Intelligence, 2020. **2**(5): p. 254-265.

[6] Kim, H., J. Na, and W.B. Lee, *Generative chemical transformer: neural machine learning of molecular geometric structures from chemical language via attention.* Journal of chemical information and modeling, 2021. **61**(12): p. 5804-5814.

361i Integrated Computational and Experimental Design of Selective Lanthanide Binding Peptides

Yiming Wang, Jason Marmorstein, E. James Petersson, Ivan J. Dmochowski, Ravi Radhakrishnan, Kathleen J. Stebe

Rare earth elements (REE), or lanthanides, are critical materials with a wide range of applications in clean energy industrysuch as battery, semiconductor, and electric vehicles, due to their unique luminescent, magnetic, and catalytic properties. Therefore, there is a growing demand for developing green industrial-scale extraction of lanthanides. However, efficient separation of heavy lanthanides (e.g. from Europium to Lutetium) with exceedingly similar chemical properties is technologically challenging, and is an active area of research. Here, we combine molecular dynamics (MD), enhanced sampling, and machine learning methods to design insilico short 17-residue peptides that selectively bind to trivalent REE cations in aqueous solution. The REE-peptide binding affinity are predicted by ML models that are trained on features that characterize the REE-peptide binding complex including hydration properties of REEs from literature studies, REE-peptide interaction energies obtained from MD simulations, and physicochemical properties of the peptide estimated from the DBAASP peptide database. Starting from the original lanthanide-binding tag (LBT) peptide, we computationally screen peptide candidates with mutations mainly on the six key ligating residues that improve predicted binding selectivity against 12 different REEs. Candidates with top selectivity scores selected from in silico discovery are verified by the fluorescence titration experiments. Moreover, we utilize a novel genetic algorithm framework to iteratively guide the design of new peptide sequences combining our model and experiments. Therefore, we demonstrate an integrated data-driven approach for quantitative sequence-structure-activity relationship as well as design of lanthanide binding peptides, paving the road for designing novel biomolecules for efficient rare earth elements recovery.

361j Integrating Automation and Machine Learning for Accelerated Closed-Loop Molecular Discovery

Brent A. Koscher, Richard B. Canty, Matthew A. McDonald, Klavs F. Jensen

The discovery of functional materials has traditionally been driven by chemists' intuition and manual experimentation; the automation of the design-make-test-analyze (DMTA) cycle presents a promising alternative for accelerating multi-functional materials discovery.

Integrating advancements in automatization and machine learning (ML), we developed a closed-loop, autonomous molecular discovery platform to both accelerate hit-to-lead (exploration) and lead-optimization (exploitation) workflows for molecules with multiple desired properties. The platform was designed for general chemistry with general, adaptable synthesis, workup, isolation, and characterization protocols. These capabilities can accommodate the diversity of molecules and multi-step syntheses recommended by its molecular generation and retrosynthesis algorithms. We demonstrate this platform in exploration and exploitation case studies without human intervention (beyond providing consumables and error recovery) on dye-like molecules targeting wavelength of maximum absorption, water–octanol partitioning, and photo-oxidative stability.

In the exploration study, the platform, guided by property-prediction ML models, automatically performed three iterations of the DMTA cycle. Using inexpensive syntheses to diverse targets as its means to explore and learn the structure–function space of four rarely reported scaffolds, the platform realized 312 unreported molecules and substantially improved the predictions of the guiding property-prediction models in these local chemical spaces. In the exploitation study, the same platform leveraged models trained on pre-existing examples of a fifth scaffold to discover 6 top-performing molecules within their structure–property space.

The platform demonstrates the potential for automated platforms which integrate the steps of the molecular discovery cycle (prediction, synthesis, measurement, and model retraining) to autonomously understand a local chemical space to aid in the discovery of functional molecules. As the components of this automated DMTA cycle are modular, with alternate property or retrosynthesis models, additional functionalities could be explored, such as biological activity, requiring only the deployment of the corresponding characterization hardware and/or protocols.

This platform is the culmination of many collaborators across multiple laboratories, and as such, this presentation will focus on the design of the experimental platform and its integration with its governing models to achieve autonomous operation.

88a Accelerating discovery with computational chemistry in challenging materials spaces

Heather Kulik

Over the past 15 years, computational chemistry modeling with first-principles density functional theory has been transformed by advances in algorithms and computing power. More recently, acceleration via machine learning has led to orders of magnitude changes in scales of materials that can be studied. Nevertheless, challenges remain in identifying which first-principles model is sufficiently accurate for property prediction needed for the accelerated screening of materials. In few areas is this more challenging than in the open shell first-row transition metals that are ubiquitous in catalysis and functional materials. I will describe our efforts to build machine learning models that overcome limitations in first-principles modeling by either building models that guide method selection or leveraging consensus from functionals to build more robust workflows for the discovery of open shell transition metal complexes with light-harvesting and catalysis applications. I will also describe how we bypass direct simulation by building machine learning models to predict experimental measures of stability with application to metal-organic frameworks and use these models to guide de novo materials selection. Beyond these advances, I will describe challenges that remain in computationally guided materials discovery.

88b Computational and Experimental Characterization of the Ligand Environment of a Ni-Oxo Catalyst Supported in Metal–Organic Framework NU-1000

Stephen Vicchio, Zhihengyu Chen, Karena Chapman, Rachel Getman

Heterogeneous catalysts exhibit significant changes in composition under different operating conditions, and these compositional changes can have significant effects on catalytic performance. For bulk metal heterogeneous catalysts, relationships between composition and catalytic operating conditions are well documented. However, the influence of operating conditions on the compositions of ligated metal cation catalysts remains largely unresolved. To address this, we use a combination of density functional theory (DFT), first principles thermodynamics (FPT), and pair distribution function (PDF) analysis to investigate ligand environments present on a $Ni_4O_xH_y$ cluster supported in the metal-organic framework NU-1000. Experimental PDFs provide insights into both the Ni-O coordination numbers and the Ni–O, Ni···Ni, and Ni···Zr distances, while DFT and FPT provide molecular level insights into the ligand environments that could result in the PDFs observed experimentally. A total of 854 ligand environments are investigated. Our findings suggest significant OH and H_2O content, and that different Ni ions within the cluster and/or NU-1000 structure may comprise subtly different numbers of these ligands. These insights will be useful for modeling catalysis on Ni oxo and other similar catalysts.

88d Electronic Structure and Reactivity Predictions for Soft Material Length Scales

Nicholas Jackson

Fundamental knowledge gaps are endemic in our understanding of how emergent properties of soft materials are linked to the quantum mechanical (QM) world. The limitations of current modeling paradigms inhibit the understanding and design of classes of soft materials for which QM phenomenology is critical. I will describe some of our recent work accelerating QM predictions of reactivity in simulations of soft materials, allowing access to traditionally difficult mechanical predictions under extreme deformation. I will also discuss the prospects for mapping rigorous QM predictions to traditional coarse-grained models from the 'bottom-up', highlighting the potential for translating the power of black-box quantum chemistry calculations to coarse-grained resolutions of soft materials.

88e Understanding the Effect of Water on CO_2 Capture in Metal-Organic Frameworks

Maytham Alzayer, Filip Formalik, Faramarz Joodaki, Srinivasu Kancharlapalli, Jiayang Liu, Randall Snurr

Carbon capture is likely to play an important role in meeting global emissions targets, and metal-organic frameworks (MOFs) are a promising class of materials that can be finely tailored for such separations. While many studies have investigated MOFs for CO_2/N_2 separations, far fewer have considered the important effect of water on CO_2 capture performance. Here, we used a variety of tools from molecular modeling and data science to find top-performing MOFs for CO_2 capture in the presence of humidity and to better understand the structure/performance relationships. In one strategy, we used a multi-scale modeling approach to screen for materials that preferentially adsorb CO_2 over both H_2O and N_2. In another strategy, we created on the computer variations of existing materials with good performance to see if we could improve their behavior, especially at high relative humidity. In a third approach, we postulated functional groups that could be incorporated into stable zirconium MOFs to improve their CO_2/N_2 selectivity in the presence of water. Results from these three strategies will be compared, enabling overarching conclusions for future directions.

294a Molecular Simulations to Understand Adsorption and Transport in Complex Zeolite Nanosheet Membranes for Ammonia/Nitrogen/Hydrogen Separation

Prerna Prerna, Jesse Prelesnik, Roshan Ashokbhai Patel, Michael Tsapatsis, J. Ilja Siepmann

Ammonia is an industrially relevant chemical with uses in fertilizer synthesis and energy storage. Conventionally, ammonia is produced through the catalytic Haber–Bosh process at pressures of 100–250 bar and temperatures of 650–750 K, followed by separation that occurs at much lower temperatures. However, the increasing demand coupled with the drive toward small-scale distributed manufacturing of ammonia requires a facile and energy-efficient ammonia separation from the stream of unreacted gases without cooling down the gas mixture. Hence, we want to explore the utilization of membrane–based separation at process–relevant conditions for effecting this separation. Our previous work has shown substantial ammonia selectivity at the ambient[1] and process–relevant conditions[2] for bulk MFI zeolite and MFI nanosheets (NS) with silanols at the surface. However, experimentally synthesized membranes are composed of multiple nanosheets layered on top of each other with non-uniform interfacial-sheet regions. Simulations can provide molecular-level insights and predictions for the separation performance, which is challenging to study via laboratory experiments because of high temperature and pressure conditions relevant for the industrial ammonia production process.

Here, we investigate complex adsorbents formed by stacking MFI nanosheets, referred to as stacked nanosheets (SNS) to understand the influence of multi–layer nanosheets on gas adsorption, separation, and diffusion. These SNS are generated through reactive force-field-based molecular dynamics simulations. Realistically, the nanosheets do not perfectly overlap with each other in the interfacial region and modeling these SNS containing misaligned channels can provide further insights into their effect on relative transport behavior of NH_3, N_2, and H_2.

We performed *NpT*-Gibbs ensemble Monte Carlo (GEMC) simulations to understand the mixture adsorption behavior of SNS models, which show that NH_3 selectively adsorbs over N_2 or H_2 in the interfacial-sheet region of SNS with silanols, and its enrichment in this region consequently affects the transport properties. Considering all these possibilities to probe SNS models with varying translational and rotational shifts, along with various mixture compositions require a large number of simulations. To predict mixture adsorption isotherms, we tested the performance of ideal adsorption solution theory (IAST), however, significant deviations were observed between adsorption loadings from IAST and *NpT*-GEMC simulations. This could be attributed to the complex structure as well as difference in adsorption behavior of NH_3, N_2, and H_2. The effect on NH_3 selectivity with varying $NH_3/N_2/H_2$ gas mixture composition is an important aspect of the ongoing work, allowing extension to process-scale modeling and optimization.

[1] Duan, X.; Kim, D.; Narasimharao, K.; Al-Thabaiti, S.; Tsapatsis, M. High-Performance Ammonia-Selective MFI Nanosheet Membranes. Chemical Communications 2021, 57, 580–582.

[2] Patel, R.; Castro, J.; Tsapatsis, M.; Siepmann, J. I. Molecular Simulations Probing the Adsorption and Diffusion of Ammonia, Nitrogen, Hydrogen, and Their Mixtures in Bulk MFI Zeolite and MFI Nanosheets at High Temperature and Pressure. Journal of Chemical and Engineering Data 2022, 67, 1779–1791.

294b Insights into Hydride Formation on Graphene-Supported Metal Nanostructures through a Combined Experimental and Theoretical Approach

Simuck F. Yuk, Walter J. Cesarski, Tyler Komorowski, Barry-John Baxter, Katherine G. Lareau, Francesca Mangione, Sebastien Thomas, Alisan S. Behr, Jeffrey Chin, Rosemary L. Calabro, Stephen F. Bartolucci, Joshua A. Maurer, F. John Burpo, Caspar Yi, Enoch Nagelli

Graphene and graphene-based materials have gained considerable interest due to their unique electronic, thermal, and mechanical performance. The functionalization of these systems has already been applied in fuel cells, energy storage, and wastewater remediation. However, understanding the interaction between metal nanostructures and graphene remains crucial to fully utilize the properties of graphene-based materials. In this study, we explored multiple reduction pathways of noble metal clusters (consisting of 16 metal atoms) supported on graphene oxide surface during chemical reduction. Density functional theory (DFT) calculations and *ab initio* molecular dynamic (AIMD) simulations were conducted in combination with experimental Fourier transform infrared (FT-IR) and X-ray diffraction analysis (XRD) measurements for understanding the surface characteristics of graphene-supported metal nanostructures.

Vibrational density of states (VDOS) was collected through simulating AIMD trajectories to study the critical surface chemistry features, including O-H stretching region at 3200-4000 cm^{-1}, presented in the experimental FT-IR spectra over the graphene oxide surface. An absence of hydride phase is also shown on Au cluster compared to Pd and Pt clusters from the XRD profiles. Thus, three mechanistic pathways were evaluated (in terms of DFT-based Gibbs free energy) for hydride formation on the surface of Au, Pd, and Pt clusters: 1) direct transfer of H^+ from the surface OH group, 2) H_3O^+-mediated H^+ transfer, and 3) transfer of H^+ through H_2 dissociation. We found that the hydride formation over supported metal clusters is thermodynamically and kinetically driven through the dissociation of H_2 molecule on the metal surface, with Au being resistant to such reduction mechanisms. Overall, our results show that a different level of hydride formation and its effects on the catalytic activities is expected with graphene oxide surface decorated with various metals.

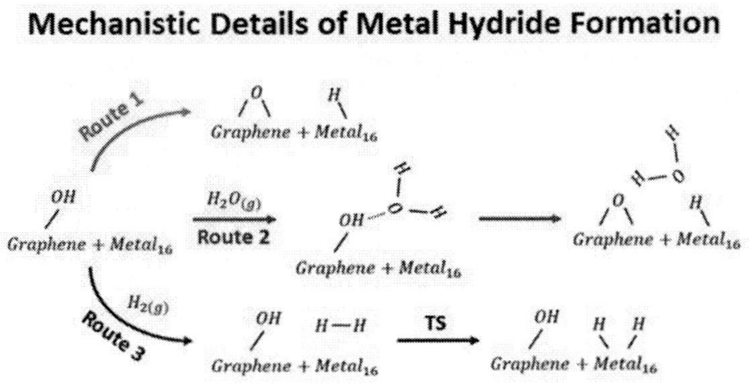

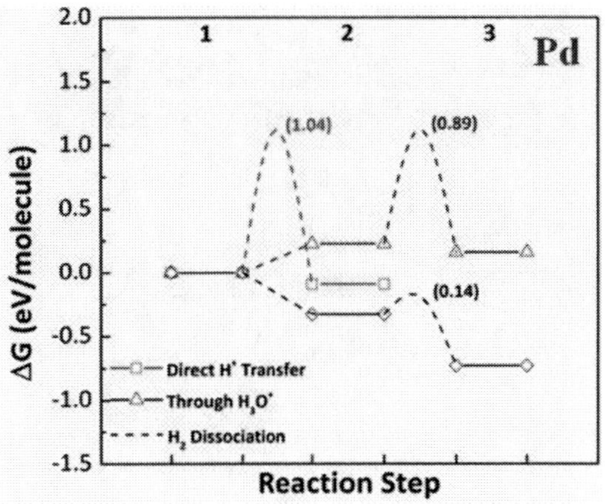

294c Atomistic Simulations of Polymer/Alumina Catalyst Support Interfacial Interactions for Applications in Plastics Upcycling

Ziyue Dong, Erin Dunphy, Michael F. Toney, Kayla G. Sprenger

Upcycling end-of-life plastics into alternative chemicals is critical to envisioning a waste-free and circular plastic economy. Hydrogenolysis, using supported metal catalysts, has shown promise as a viable catalytic pathway for transforming plastic wastes into new value-added chemicals in an economical manner. However, the complex chemical makeup of mixed plastic waste streams poses a major challenge to realizing this opportunity. Understanding the key molecular interactions that take place at the inorganic-organic depolymerization interface and directly affect catalytic performance, from a computational perspective, is thus paramount to improving the design of new catalysts for plastics hydrogenolysis.

We hypothesized that interactions between the polymers and the catalytic surface would be governed by the presence of the metal oxide support. As such, we performed physics-based, molecular dynamics simulations on the polymer-support interfacial system to elucidate these key interactions. Here, we report the interfacial molecular layering, orientational distribution, and free energies of interaction, of polyester polylactic acid and polyethylene terephthalate, as well as polyolefin polyethylene, at the alumina metal oxide support surface. Our simulation results identified variable effects of polymer chain length and functional groups on the interfacial molecular behavior of the polymers, which will be validated using X-Ray reflectivity experiments on analogous *in vitro* model systems. In this investigation, we applied state-of-the-art enhanced sampling techniques well-tempered metadynamics and parallel tempering metadynamics in the well-tempered ensemble. These methods rendered highly accurate descriptions of polymer-support interactions and near-surface polymer configurations over timescales not previously achievable using classical simulations. Molecular insights gained from this work can be leveraged to improve future catalyst designs and propose catalytic mechanisms that directly affect reactivity and selectivity. The general simulation workflow derived from this work can also be applied to elucidate interfacial phenomena at other inorganic-organic interfaces, such as that of lithium-ion battery electrodes and of organically-functionalized colloidal nanocrystals.

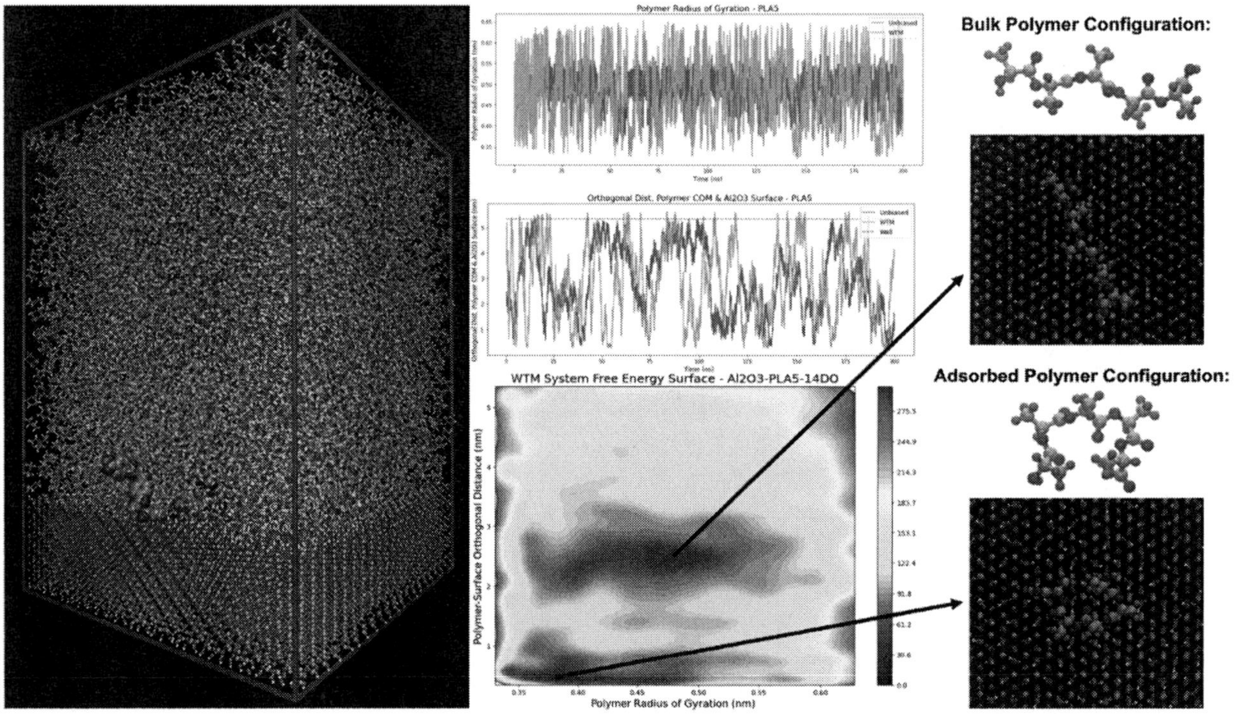

294d A Molecular Perspective of Wetting and Charging at Polymer Interfaces

Michael Webb, Hang Zhang, Sankaran Sundaresan

Interactions at water-polymer surfaces underlie many important natural and technological phenomena. In this talk, I will discuss quantitative theoretical approaches—namely molecular-level analyses and free-energy calculations—to understand and quantify the (i) wetting and (ii) contact charging behavior of chemically distinct polymer surfaces. Related to wetting, we will show how hydrogen-bonding behavior and surface morphology dictates and differentiates polymer surfaces, even when macroscopic measures of hydrophobicity fail to resolve any differences. Related to contact charging, which refers to how surfaces become asymmetrically charged following contact and separation, we will evaluate a new, molecularly informed hypothesis that can predict the direction of charge transfer between two surfaces. Although water is thought to play a central role in contact charging of insulating surfaces, like polymers, the mechanism of charging is unknown and hard to isolate via experiment. Whereas many existing hypotheses rely on kinetic arguments, we assess whether thermodynamic driving forces, informed by our free energy calculations, can feasibly explain experimental trends. Overall, this presentation will highlight how molecular simulation can help us to understand complex interfacial phenomena.

294e Understanding the Role of Polymers on the Nucleating Behavior of Water in Dilute Supercooled Solutions Using Molecular Dynamics Simulations

Aindrila Indra, Mangesh Bhendale, Jayant K. Singh

The ice recrystallization inhibition (IRI) phenomenon is of utmost importance in the field of medicine (e.g., cryopreservation of organs, drug delivery), agriculture (e.g., crop damage), and industrial applications (e.g., jet aviation fuel, de-icing of turbine blades) [1-3]. Hence, understanding the nucleation behavior of water in dilute polymeric solutions is important for the development of suitable artificial IRI agents. Although Poly (Vinyl Alcohol) (PVA) is found to be one of the most potent biomimetic IRI agents, the molecular understanding of the nucleation behavior of water in the presence of such polymers is still lacking.

Here, in this work, we use molecular dynamics to elucidate the role of the concentration, degree of supercooling, degree of polymerization, and amphiphilicity of PVA and PVA-like polymers on the homogeneous nucleation of water in dilute polymeric solutions using the seeding method, based on the Classical Nucleation Theory (CNT) [1,4,5]. The seeding simulations are used to determine the critical cluster size (N_c), which has an equal probability to either grow or melt. The predicted N_c is then used to evaluate the interfacial free energy, the nucleation free energy barrier and then finally the nucleation rate of the supercooled water, based on the CNT formalism.

In line with the literature [1,2], our results indicate that the homogeneous nucleation rate of water in dilute supercooled polymeric solutions decreases with an increase in the solute concentration. The increase in the chemical potential difference and the interfacial free energy between ice and the supercooled solution, with an increase in the solute concentration, led to an increase in the nucleation free energy barrier, which finally decreased the nucleation rate of water. The nucleation rate also decreased with a decrease in the degree of supercooling due to the reduced propensity of the water molecules to nucleate at higher temperatures. The effect of the chain length and amphiphilicity of the polymers on the inhibition of ice nucleation is dominant at a higher degree of supercooling, whereas, at lower supercooling, the inhibition is mainly due to an increase in the solute concentration.

Our simulation results provide preliminary molecular insights into the factors affecting the nucleating tendency of water in presence of amphiphilic polymers, which may play a pivotal role in designing artificial IRI agents, for use in cryobiology, drug delivery and tissue engineering applications.

Figure 1: (a) The initial configuration for seeding simulations with Ih seed (red), PVA (black), and supercooled liquid water (cyan), (b) the time evolution of the number of ice molecules in the largest seed in the simulation box. The image represents the critical seed size of (N_c=2818) molecules for 2 wt% PVA 4-mer solution at 252.5 K, and c) the nucleation rate of water in supercooled solution at 240 K. The solid and empty symbols represent the data for solutions with 4-mer and 8-mer polymer molecules, respectively.

References

[1] Mochizuki K. et. Al., *J. Am. Chem. Soc.* 139, 17003–17006 (2017)

[2] Naullage P.M. and Mollinero V., *J. Am. Chem. Soc.* 142, 4356–4366 (2020)

[3] Naullage P.M. et. al., *J. Phys. Chem.* 121, 26949–26957 (2017)

[4] Espinosa J. R. et. al., *J. Chem. Phys.* 141, 18C529 (2014)

[5] Goswami A. et. al., *J. Chem. Phys.* 153, 094502 (2020)

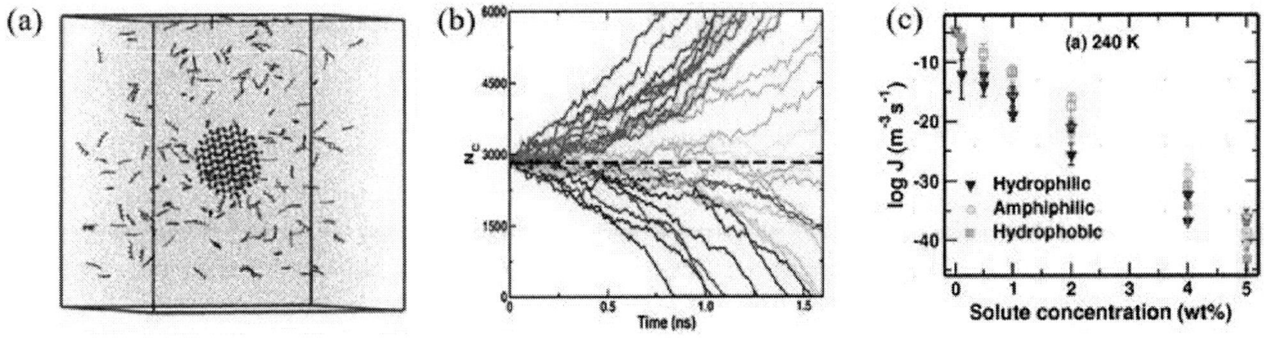

(a)
(b)
6000
4500
3000
1500
0
N_c
0
0.5
1.0
1.5
Time (ns)
(c)
(a) 240 K
-10
-20
-30
-40
log J (m^{-3} s^{-1})
Hydrophilic
Amphiphilic
Hydrophobic
0
1
2
3
4
5
Solute concentration (wt%)

294f Computer-Aided Molecular Design of Heat-Resistant Surfactants Using Classical Density Functional Theory

Pierre Walker, Sriteja Mantha, Yasemin Basdogan, Zhen-Gang Wang

The investigation of innovative surfactants is a highly active research area, particularly regarding the identification of surfactants suitable for specific applications such as enhanced oil recovery. However, due to the scarcity of large-scale experimental databases, the implementation of machine-learning tools for the design of such molecules is a challenging task. Nonetheless, predictive models, such as classical density functional theory (cDFT), can be used in combination with a group-contribution methodology to overcome this problem, enabling the use of computer-aided molecular design (CAMD) frameworks. Therefore, in this study, we have developed open-source, generalized frameworks for both cDFT and CAMD. By implementing these tools, we utilized the well-established heterosegmented PC-SAFT functional coupled with mixed-integer non-linear optimization algorithms to design heat-resistant surfactants. The proposed molecules aligned with our anticipated structures for such surfactants, some of which have yet to be examined experimentally or through molecular dynamics simulations. Our intention is to expand our current framework to design anionic, cationic, and zwitterionic surfactants. Furthermore, these frameworks can also be extended to explore other design problems such as electrolytes for battery applications and solvents for carbon capture absorption.

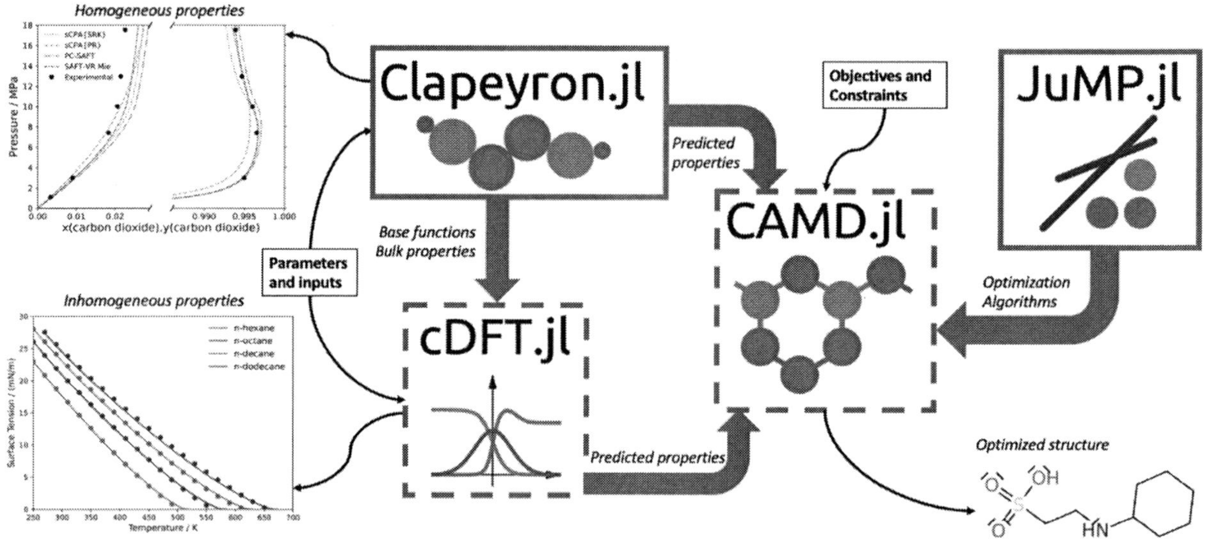

294g Enhanced Stabilization of Cytokines for Nanoparticle Delivery Applications

Emily R. Rhodes, Nicole B. Day, Emma C. Aldrich, C. Wyatt Shields IV, Kayla G. Sprenger

Cytokines are small proteins that are crucial to cell signaling, making them useful as therapeutic agents. However, cytokines must be localized to specific areas of the body to minimize harmful side effects such as systemic inflammation or cytokine storm, requiring targeted drug delivery methods. Polymeric micro- and nanoparticles have emerged in recent years as promising vehicles for improving the delivery of cytokines to diseased tissues (e.g., solid tumors).[1,2] However, cytokines are also sensitive to denaturation, which may limit the number of methods available to formulate micro-/nanoparticles, depending on the cytokines being delivered. A clear molecular understanding of the interactions between the relevant proteins and solvents used in these systems could enable the improved design of cytokine-encapsulating particles for delivery to, and function within, various tissue microenvironments.

We utilized an integrated computational and experimental approach for studying the governing mechanisms by which a representative cytokine (interleukin-12; IL-12) is protected from denaturation by a protein stabilizer (bovine serum albumin; BSA) at an organic solvent-water interface found in some nanoparticular formulations. Specifically, we investigated key protein-protein interactions between IL-12 and BSA. We simulated these components both in water and in the presence of a water/dichloromethane (DCM) interface to mimic our double emulsion formulation method. Briefly, we observed that: (i) IL-12 partially denatures near the water/DCM interface, and (ii) IL-12 denaturation is reduced in the presence of BSA. Experimentally, we found that limited cytokine ($<1\%$ of loaded IL-12) can be encapsulated in particles without a protein stabilizer. Conversely, if BSA is added to the aqueous phase in molar excess to IL-12 and sufficient time is allowed for binding, elevated levels of cytokine can be encapsulated. Findings from this work have implications in establishing design principles to stabilize cytokines into polymeric particles for robust delivery to diseased tissues, enhancing their use as therapeutic agents for treating disease.

References

[1] Zhao, Z. *et al.* "Systemic tumour suppression via the preferential accumulation of erythrocyte-anchored chemokine-encapsulating nanoparticles in lung metastases", *Nat Biomed Eng*, **2021**, 5, 441–454.

[2] Zhao, Z. *et al.*, "Erythrocyte leveraged chemotherapy (ELeCt): Nanoparticle assembly on erythrocyte surface to combat lung metastasis", *Sci Adv*, **2019**, 5, eaax9250.

294h Engineering Osteogenic Peptides through Machine Learning

Ricky Dash, Esmaiel Jabbari

Human bone morphogenetic proteins (BMPs) are used in certain clinical applications such as spine fusion and non-unions to accelerate bone regeneration and healing. However, due to their short-half-life and diffusion, doses much higher than the endogenous amount are used clinically to stimulate bone formation. These high doses cause undesired side effects such as bone overgrowth and tumorigenesis, which has limited their widespread use in orthopedic surgery. An alternative approach is to use peptides derived from the bioactive domains of BMPs to reduce side effects[1, 2]. However, the osteogenic activity of these peptides is much lower than that of BMPs which is attributed to conformational and solubility differences between the native and free peptide. In this regard, there is a need to design and discover novel peptides that mimic the native conformation and solubility of the epitope sequence on BMPs that interact with cell surface receptors. The bioactive domain of the BMP-2 peptide known as the knuckle epitope has an open arm structure when it is a part of the protein but behaves differently in its free form indirectly affecting its osteogenecity[3, 4]. Hence there is a need to come up with new peptide sequences which mimic the knuckle epitope sequence and has biological activity comparable to that of the BMP-2 protein. We hypothesize that such biomimetic peptides can be engineered using the power of mesoscale simulation and machine learning. We developed a uniquely parameterized mesoscale simulation method or Dissipative particle dynamics method, in which the macromolecules are coarse-grained into beads, to determine the biophysical properties of the peptides. The forcefield for this method included evaluating the conservative potential term by considering specific interactions between the different beads and modelling the bonded interaction potentials such as the bond stretching as harmonic bonds, bond angles assume cosine harmonic functional forms and dihedral angle energies are modelled through a shifted dihedral potential[5]. The electrostatic interactions were determined using a function that allows to distribute or smear charges on the bead. We examined this method at different coarse-graining levels or mappings for a known 12-mer peptide TrpZip2 (PDB ID: 1LE1). A mesostructure was created with one peptide in a box of water with periodic boundary conditions and simulated under NVT conditions. We compared the radius of gyration values of the simulated backbone with the actual structure. The radius of gyration of the 20 structures in the PDB database was in the range 5.6 A - 6.2 A. The average of the radius of gyration of the simulated peptide structure was within the range of the radius of gyration values of the actual structure. This method can be used to predict the structural properties of peptides in a quick and reliable way. Using this method and existing experimental characterization techniques, the structural properties and biological activity of several modified knuckle epitope sequences will be determined to create a database. This database will be used to train and build a Quantitative Structure-Activity relationship model through machine learning algorithms which can then generate potential osteogenic peptide sequences. The discovery of highly active peptides could potentially expand their use in orthopedic applications to improve the quality of life of patients with skeletal injuries.

References

[1] R. Visser, G.A. Rico-Llanos, H. Pulkkinen, J. Becerra, Peptides for bone tissue engineering, Journal of Controlled Release 244 (2016) 122-135.

[2] G. Tang, Z. Liu, Y. Liu, J. Yu, X. Wang, Z. Tan, X. Ye, Recent Trends in the Development of Bone Regenerative Biomaterials, Frontiers in Cell and Developmental Biology 9 (2021) 665813.

[3] G. Zhao, L. Zhang, L. Che, H. Li, Y. Liu, J. Fang, Revisiting bone morphogenetic protein-2 knuckle epitope and redesigning the epitope-derived peptides, Journal of Peptide Science 27(6) (2021).

[4] S. Moeinzadeh, D. Barati, S.K. Sarvestani, T. Karimi, E. Jabbari, Experimental and Computational Investigation of the Effect of Hydrophobicity on Aggregation and Osteoinductive Potential of BMP-2-Derived Peptide in a Hydrogel Matrix, Tissue Engineering Part A 21(1-2) (2015) 134-146.

[5] R. Vaiwala, K.G. Ayappa, A generic force field for simulating native protein structures using dissipative particle dynamics, Soft Matter 17(42) (2021) 9772-9785.

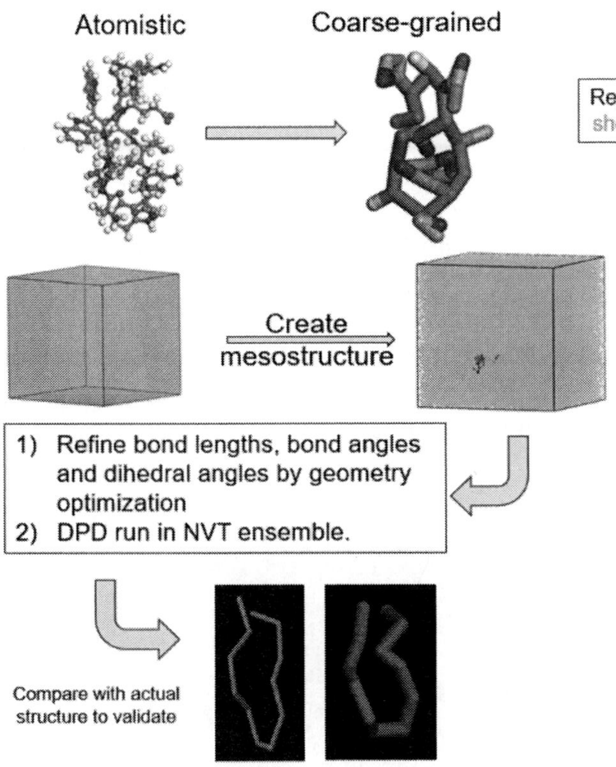

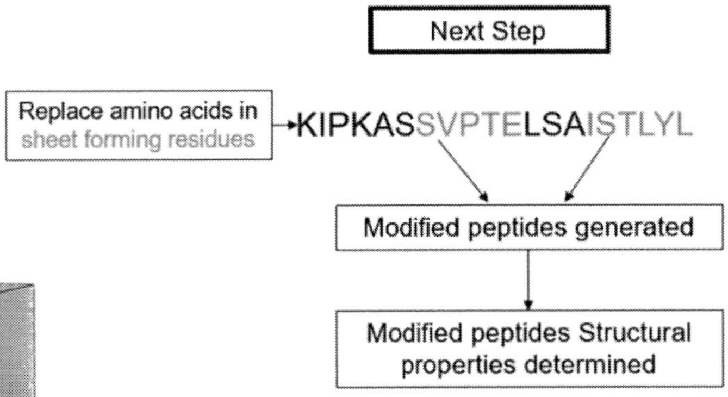

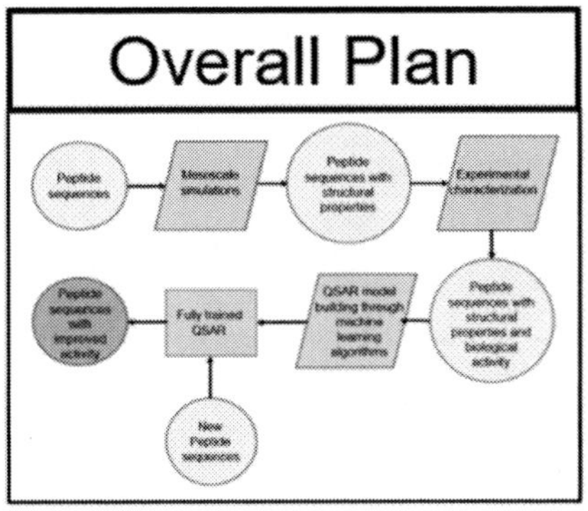

294i Dynamics of Surfactant Protein B at the Alveolar Air-Liquid Interface: Insights from Molecular Modeling and Simulations

Mona Minkara, Tyler Locke

In this study, we present a computational model developed to investigate the dynamics of surfactant protein B (SP-B), a crucial component of pulmonary surfactant that plays a vital role in transitioning the pulmonary surfactant lipid layer during inhalation and exhalation. To model SP-B, we used homology modeling based on the crystallized Mini B structure and the knowledge that SP-B is part of the saposin family of proteins. Molecular dynamics simulations were conducted on both open and closed states of SP-B in hydrophilic and hydrophobic environments to simulate the conditions found within the alveoli. We then observed the resulting structural changes. Our findings provide insight into the behavior of SP-B in different conditions and may help inform the development of more effective therapies for pulmonary diseases associated with surfactant dysfunction.

294j Insights into the Mass Transfer through Vapor-Liquid Interfaces from Molecular Dynamics Simulations

Simon Stephan, Dominik Schaefer, Hans Hasse

In fluid separation processes, both thermodynamic bulk and interfacial properties are important. In distillation and absorption processes, the mass transfer through vapor-liquid interfaces plays a central role. In classical theory, the vapor-liquid interface is a two-dimensional object. In reality it is a region in which properties change smoothly over a few nanometers from its liquid bulk to its gas bulk value. Properties of that nanoscopic interfacial region can presently not be measured experimentally, but can be predicted using methods from molecular thermodynamics. Different theoretical methods consistently predict an enrichment of low-boiling components in many mixtures [1], which is suspected to influence the mass transfer through the interface. This enrichment would be an insurmountable obstacle to mass transfer according to the Fickian theory and, hence, is suspected to influence for example fluid separation processes like absorption or distillation.

To study the influence of the enrichment on the mass transfer, two new simulation methods based non-equilibrium molecular dynamics simulation (NEMD) were developed by our group [2,3]. The first method establishes a stationary mass flux across vapor-liquid interfaces and the second method investigates the transient mass flux of a low-boiling component through an interface into a liquid phase. For the stationary method, averaging is carried out using the classical block averaging method. In the non-stationary method, averaging is carried out using the replica method. Two model systems were used for testing and comparing the new NEMD simulation methods. The non-stationary method is found to be significantly more robust and insightful compared to the stationary method. In particular, the non-stationary method provides a high-resolution insight into the response functions for the component densities, fluxes, and pressure in the interfacial region. The benefits and drawbacks of the two methods will be discussed in detail in this contribution.

To study the influence of the nature of the interface, i.e. enrichment, surface tension, thickness etc., on the mass transfer through the vapor-liquid interface, a comprehensive simulation study was carried out focusing on the non-stationary simulation method. The influence of the temperature, the (bulk phase) composition, and the mixture type on the interfacial mass transfer effects was thereby studied. Five binary Lennard-Jones model systems were considered: An ideal system (A), a high-boiling azeotropic system (B), a low-boiling azeotropic system (C), a heteroazeotropic system (D), and a system exhibiting a high pressure VLLE (E). These systems are good candidates for a systematic study as their phase equilibrium and equilibrium interfacial properties [3,4] as well as their bulk phase transport coefficients [6,7] have been systematically studied in a previous work of our group. In particular, the five systems exhibit important differences in their interfacial properties. For example, at equilibrium states, the enrichment of the low-boiling component at the interface increases going from mixture A to E. The results from the NEMD simulations indicate a significant influence of the enrichment on the mass transfer. Both simulation methods consistently yield a reduced mass flux (about an order of magnitude) in the case of high enrichment (systems D and E) compared to cases with no or low enrichment (systems A to C). It is shown that the differences in transport properties of the bulk phases alone cannot account for the findings of the NEMD simulations. Hence, the differences in the mass flux are attributed to interfacial effects. Furthermore, it was observed, that the component entering the liquid phase through an interface may be repelled from the interface. In some cases, even a net negative mass flux back into the vapor phase is observed due to the repelling of particles from the interface. This repelling of particles significantly depends on the nature of the interface, i.e. its surface tension and interfacial thickness.

References

[1] S. Stephan, H. Hasse, Int. Rev. Phys. Chem. 39, 3 (2020) 319-349.

[2] S. Stephan et al., Mol. Phys. 119, 3 (2021) e1810798.
[3] D. Schäfer, S. Stephan et al., J. Phys. Chem. B (2023) in press.
[4] S. Stephan, H. Hasse, Phys. Chem. Chem. Phys. 22 (2020) 12544.
[5] J. Staubach, S. Stephan, J. Chem. Phys. 157 (2022) 124702.
[6] D. Fertig, H. Hasse, S. Stephan, J. Mol. Liqu. 367 (2022) 120401.
[7] D. Fertig, S. Stephan, Mol. Phys. (2023) e2162993.

Properties of Fluid Interfaces: Insights from Molecular Thermodynamics

Simon Stephan

Fluid interfacial properties play an important role, for example in separation processes in chemical engineering, in nucleation in propulsion systems, and also in nature. The fluid interface is a region in which properties change smoothly over a few nanometers from its liquid bulk to its gas bulk value. Fluid interfacial properties are complex and experimentally hard to examine due to the small length scale and fluctuations present at the interface (both spatial and temporal). In this contribution, results from recent studies from our group [1-11] on equilibrium properties of fluid interfaces of molecular fluids are presented that were obtained using methods based on molecular thermodynamics, namely molecular dynamics (MD) simulations, density gradient theory (DGT) + molecular-based equations of state, and (in some cases) experiments. Different types of fluid interfaces are considered, i.e. vapor-liquid, liquid-liquid, and vapor-liquid-liquid interfaces. The focus is on vapor-liquid interfaces. Both model fluids and real substance fluids were studied. The model fluids were used to study the effect of the influence of molecular interaction parameters on interfacial properties; the real substances were chosen as they are applicable for industrial processes.

This work contributes on the understanding of the fundamentals of fluid interfacial properties and progresses modelling methods. Multiple features and their interrelations are addressed: (i) general features of the monotonicity behaviour of profiles of thermodynamic properties across fluid interfaces are discussed [1] by critically revisiting statements made in the seminal monograph of Rowlinson and Widom 'Molecular Theory of Capillarity'. It is thereby shown that two non-monotonicity features can be distinguished: (a) the enrichment of components at the interface of mixtures [1,2] and (b) the oscillatory layering structure (OLS) of particles at fluid interfaces [1,3]; (ii) the enrichment, which is suspected to influence the mass transfer across interfaces [4,5], is systematically studied regarding its dependency on temperature and composition in various binary mixtures [2,6,7]; (iii) a new method is presented for determining the OLS. The results are used to confirm a link between the OLS and the Fisher-Widom line. The results are found to be in good agreement with results from the literature; (iv) the influence of long-range interactions on the monotonicity behaviour and other mixture interfacial properties is discussed [8]; (v) it is shown that conformal solution theory can be extended and applied for the modelling properties of fluid interfaces of mixtures – including the enrichment [9]. It is thereby shown that interfacial properties are directly linked to the mean interactions of the liquid phase [9]; (vi) the relation of the enrichment and the relative (Gibbs) adsorption is discussed pointing out important differences of these two closely related properties [1,2,10]. For the latter, results from both MD, DGT, and pendant drop experiments are presented [10]; (vii) the relation of interfacial properties and bulk phase equilibrium properties will be discussed using a new type of plot that combines the global phase diagram (critical lines, VLLE lines, etc.) and iso-lines for interfacial properties, which condenses the phase and interface information of a binary mixture and provides new insights into their intimate relations; (viii) it is shown that the enrichment of components at fluid interfaces can be understood as a prewetting, where a third fluid phase heterogeneously nucleates at the interface, which starts already at conditions alongside the actual three-phase equilibrium line [7,11]; (ix) an empirical enrichment model is presented that was developed using model fluid data alone using only bulk phase property descriptors. It is shown that this model can reliably predict the enrichment also for complex real substance mixtures (appr. 2000 data point from this work and retrieved from about 100 publications from the literature) [2]; (x) the relation of intersection points of profiles in the interfacial region and critical points is discussed, which can be related to the rectilinear diameter [3,6,7,11].

This work contributes to a large variety of properties and interfacial equilibrium effects of fluid interfaces. These interfacial properties are a direct result of the large gradients of the profiles of thermodynamic properties across the interface. These profiles connect the bulk values of the properties. It is therefore obvious that the interfacial properties are closely related to the respective bulk phase properties, which is demonstrated here by multiple examples (i)-(x). The properties and interfacial effects discussed in this work contribute to a picture in which the properties of the interfacial region mirror bulk properties of the system, albeit in the way of a fun-house mirror in which the bulk behavior is distorted by the strong gradients. Moreover, the interrelations of these interfacial properties and effects will be discussed as well as the question, which physical model capture which of these effects accurately. These interrelations, e.g. between the enrichment and the wetting transition as well as between different non-monotonicity effects, contribute to a better understanding of the nature of fluid interfacial properties and their relevance for different applications and conditions. Also, important new insights regarding (to date unknown and overlooked) artifacts in modelling approaches are addressed. All this demonstrates the complexity and richness of fluid interfacial properties. The complementary modelling approach using MD and DGT (and pendant drop experiments in some cases), however, provides a wealth of novel insights. Moreover, open research questions in the field of fluid interfacial properties and their modelling that should be addressed in future work are pointed out.

References

[1] S. Stephan et al.: Fluid Phase Equilib. 564 (2023) 113596.
[2] S. Stephan and H. Hasse, Int. Rev. Phys. Chem. 39, 3 (2020) 319-349.
[3] S. Stephan et al.: J. Phys. Chem. C 122 (2018) 24705.
[4] S. Stephan et al., Mol. Phys. 119, 3 (2021) e1810798.
[5] S. Stephan et al.: Chem. Eng. Trans. 69 (2018) 295.
[6] S. Stephan et al., J. Chem. Phys. 150 (2019) 174704.
[7] J. Staubach and S. Stephan, J. Chem. Phys. 157 (2022) 124702.
[8] S. Stephan and H. Hasse, Mol. Phys. 187 (2020) 1-14.
[9] S. Stephan and H. Hasse, Phys. Rev. E 101 (2020) 012802.
[10] S. Stephan et al., Fluid Phase Equilibr., 518 (2020)112583.
[11] S. Stephan and H. Hasse, Phys. Chem. Chem. Phys. 22 (2020) 12544.

Phase Behavior of Polydisperse Rod-like Particles in Solution: Depletion Interactions and Isotropic-Anisotropic Transitions

Elizabeth Adeogun, Karthik Nayani

This study investigates the phase behavior of polydisperse rod-like particles in solution using the Asakura and Oosawa (AO) depletion theory for rod-like macromolecules. The effect of depletion interactions between rods of different lengths on the ordering of rods is quantitatively evaluated. Numerical calculations are performed with different sizes of ellipsoids, and simulations are carried out using Gay-Berne particles in the NPT isothermal isobaric ensembles. The main objectives of this study are to (i) examine the depletion interaction between rod-like particles of different aspect ratios and (ii) analyze the isotropic–anisotropic phase transitions within the coexistence range. The results demonstrate that the properties of the anisotropic phase vary significantly with temperature, aspect ratio, and density, which is crucial for understanding the behavior of rod-like particles in solution.

525a Machine Learning Guided Adaptive Sampling Methods for Learning Conformational Ensemble of Proteins

Diwakar Shukla

Rapid computational exploration of the free energy landscape of biological molecules remains an active area of research due to the difficulty of sampling rare state transitions in Molecular Dynamics (MD) simulations. In recent years, an increasing number of studies have exploited Machine Learning (ML) models to enhance and analyze MD simulations. Notably, unsupervised models that extract kinetic information from a set of parallel trajectories have been proposed, including the variational approach for Markov processes (VAMP), VAMPNets, and time-lagged variational autoencoders (TVAE). In this work, we present a combination of adaptive sampling with active learning of kinetic models to accelerate the discovery of the conformational landscape of biomolecules. In particular, we introduce and compare several techniques that combine kinetic models with two adaptive sampling regimes (least counts and multi-agent reinforcement learning-based adaptive sampling) to enhance the exploration of conformational ensembles without introducing biasing forces. Moreover, inspired by the active learning approach of uncertainty-based sampling, we also present MaxEnt VAMPNet. This technique consists of restarting simulations from the microstates that maximize the Shannon entropy of a VAMPNet trained to perform soft discretization of metastable states. By running simulations on two test systems, the WLALL pentapeptide and the villin headpiece subdomain, we empirically demonstrate that MaxEnt VAMPNet results in faster exploration of conformational landscapes compared to the baseline and other proposed methods.

525b Sampling the Conformational Space of Proteins Using Anisotropic Network Model Modes Driven Metadynamics

Betul Uralcan

Complex biological macromolecules such as proteins carry within their functional, native structures the key to many essential life processes. While molecular simulations are used to explore protein conformational changes at the atomistic scale, they are often limited by accessible temporal scales. To overcome this limitation, enhanced sampling methods have been used. Parallel-tempering metadynamics (PTMetaD) is an approach that facilitates the efficient sampling of the free energy landscape based on the biasing of a set of collective variables and high temperatures that accelerate dynamics. A significant bottleneck in the success of PTMetaD is the its proneness to be affected from the choice of collective variables (CVs). In this work, we propose a new framework to rationally design efficient CVs for PTMetaD that minimize large hysteresis effects. The algorithm involves using the slow modes calculated from α-carbon-based anisotropic network model (ANM) as CVs to rationally drive the system along the highest collective modes of motion. By only using the intrinsic dynamic modes of motion that are accessible to the protein to bias simulations, convergence of the free energy landscape improves significantly compared to PTMetaD based on conventional CVs. Through simulations of multiple proteins that exhibit a diverse set of biological functions and conformational changes, we show how CVs derived from ANM modes can facilitate efficient and reliable characterization of conformational changes.

525c Comparison of on-the-Fly Probability Enhanced Sampling and Parallel Tempering Combined with Metadynamics for Atomistic Simulations of RNA Tetraloop Folding

Kosar Rahimi, Pablo M. Piaggi, Gul Zerze

Atomistic simulations with reliable models are extremely useful in providing exquisitely detailed pictures of biomolecular phenomena that are not always accessible to experiments. One such biomolecular phenomenon is RNA folding, which often requires exhaustive simulations with combined advanced sampling techniques. In this work, we employed multithermal-multiumbrella on-the-fly probability enhanced sampling (MM- OPES) technique, and compared it against combined parallel tempering and metadynamics simulations. We found that MM-OPES simulations were successful in reproducing the free energy surfaces from combined parallel tempering and metadynamics simulations. Importantly, we also investigated a wide range of temperature sets (mini- mum and maximum temperatures) for MM-OPES simulations in order to identify some guidelines for deciding the temperature limits for an accurate and efficient exploration of the free energy landscapes. We found that most temperature sets yielded almost the same accuracy in reproducing the free energy surface at the ambient conditions as long as i) the maximum temperature is reasonably high, ii) the temperature at which we run the simulation is reasonably high (in our simulations, running temperature is defined as [minimum temperature + maximum temperature]/2), and iii) effective sample size at the temperature of interest is statistically reasonable. In terms of the computational cost, all MM-OPES simulations were nearly four times less costly than the combined parallel tempering and metadynamics simulations. We concluded that the demanding combined parallel tempering and metadynamics simulations can safely be replaced with approximately four times less costly MM-OPES simulations (with carefully selected temperature limits) to obtain the same information.

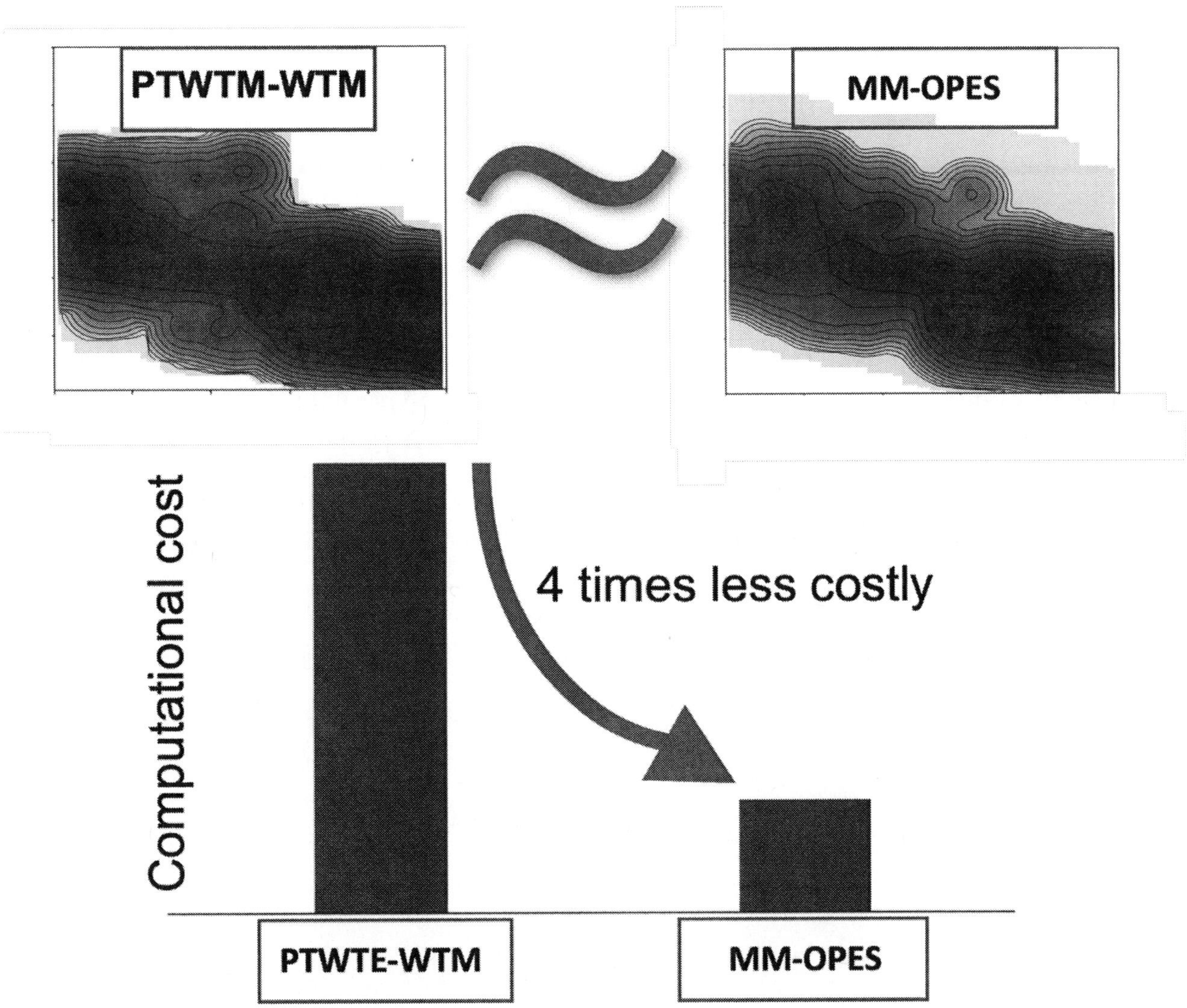
PTWTM-WTM
MM-OPES
Computational cost
4 times less costly
PTWTE-WTM
MM-OPES

525d Surrogate Modelling of Anisotropic Pair Potentials Using Smolyak Sparse Grids

Mohammadreza Fakhraei, Michael Howard, Christopher Kieslich

Mathematically representing the effective pairwise interaction between complex anisotropic bodies is challenging because the interaction can depend on many coordinates and the functional form of the interaction is generally unknown. For example, the pairwise potential of mean force and torque (PMFT) between two rigid body proteins can depend on up to 6 coordinates (relative position and orientation vectors) and is computationally demanding to evaluate. Surrogate models can represent such complex functions by sampling them for a set of inputs and training a model to approximate the function over its entire domain. Here, we develop a surrogate modelling approach to approximate anisotropic pair potentials from limited data using Chebyshev polynomial interpolants on Smolyak sparse grids. We test our approach on several model anisotropic potentials and a toy model of the TMV coat protein. With judicious variable transformations and regularization, good approximations of the potentials are obtained. We anticipate this methodology will be used to compute PMFTs from atomistic simulations to enable ultra-coarse-grained simulations of protein solutions at unprecedented length and time scales.

525e Efficient Simulation of Rare-Events Using Approximate Brownian Bridges

George Curtis, Vivek Narsimhan, Doraiswami Ramkrishna

The simulation of rare events is an important problem in chemical physics with numerous challenges and applications. To investigate such phenomena, we study a process known as a generalized Brownian bridge – i.e., a continuous random walk conditioned to lie in a specified region of phase space and/or end in a given region. This random process has broad applicability when one wants to control the endpoint of stochastic systems, which is often the case in fields like polymer physics and reaction systems. However, construction of a bridge requires solving a Backwards Fokker-Planck (BFP) equation which suffers from the "Curse of Dimensionality" and thus is impractical to compute on complex and high dimensional potential energy surfaces. Therefore, we propose leveraging approximate solutions in conjunction with an importance sampling scheme to correct (re-weight) any errors which occur. Specifically, we efficiently generate rare-events such as barrier-crossing trajectories and trajectories which stay within a region of phase-space for a long time by exploiting the asymptotic properties of the BFP and leveraging the systems dependence on the dominant eigenvalue at large times respectively. We see that these approximations drastically simplify the bridge construction while maintaining the statistical accuracy of its results. We will also compare the results of this methodology to other techniques and comment on applications where the Brownian Bridge methodology would be advantageous.

525f Calculating Entropy Production with Nonequilibrium Molecular Dynamics

Alex Albaugh, Rueih-Sheng Fu, Geyao Gu, Todd Gingrich

A system with net flows of mass or energy is not in equilibrium, a common feature of industrial and biological systems. These flows can come from many sources: heat baths at different temperatures, moving boundaries, imposed external torques or forces, inflows or outflows of mass, hydrodynamic flows, electric or magnetic coupling---the list is endless. To study nonequilibrium systems with molecular simulations requires the development of methods to model the particular mechanism behind the nonequilibrium behavior.

In this talk I will present a method for treating chemical nonequilibrium in molecular dynamics simulations. By introducing grand canonical Monte Carlo chemostats for reactants and products, the concentrations of species undergoing reactions can be held away from their thermodynamic equilibrium. I will also present a method for calculating the entropy produced in this nonequilibrium chemical state. Entropy production is a necessary component in thermodynamic uncertainty relationships, which bound the precision of a system based on the entropy produced. In this way I can compare how close a nonequilibrium system is to thermodynamic bounds.

I will demonstrate these methods by simulating a model synthetic molecular motor. By holding the system at a high concentration of fuel and low concentration of waste, away from equilibrium concentrations, the motor has a continuous supply of free energy. In the presence of this free energy supply the motor produces mechanical work. By calculating the precision of this work and the entropy produced in the system, I find that the motor model operates far from thermodynamic limits. These results suggest that the performance of synthetic motors can be greatly improved and demonstrate the power of the nonequilibrium simulation method and entropy production calculation.

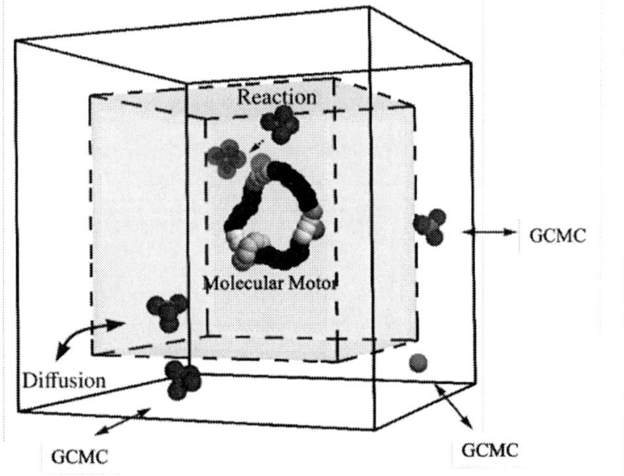

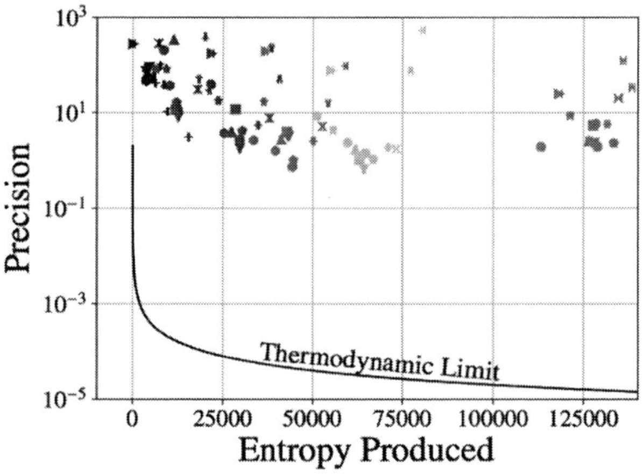

525g Non-Equilibrium Simulations As a Tool to Obtain Equilibrium Properties of Biomolecular Systems

Lev Levintov, Harish Vashisth

Free energy is an essential thermodynamic quantity which establishes the behavior of any chemical or biological system at or near equilibrium. For example, this quantity governs ligand binding-unbinding processes, diffusion of small molecules across cell membranes, conformational transitions of biomolecules and many more. Therefore, the knowledge of the associated free energy landscape is crucial for the comprehensive description of any biophysical process. However, determining a reliable free energy map is often a difficult task to perform. Non-equilibrium simulations could serve as a powerful tool for characterizing equilibrium properties of biomolecular systems, including equilibrium free energy differences. Here, we present an application of non-equilibrium constant velocity steered molecular dynamics (cv-SMD) simulations in conjunction with Jarzynski equality to probe the binding-unbinding process of ligands from viral RNA elements. Following this approach, we characterized these processes using equilibrium free energy profile and identified conformational transitions associated with them.

525h Modeling Thermosetting Polymers with Hierarchical Nanostructure

Jacob Gissinger

Block copolymers self-assemble into a variety of morphologies that can serve as atomic-scale templates for the creation of nanostructured metals or other classes of polymers, such as high-performance thermosets. These nanonetworks have unique mechanical and transport properties, allowing them to function as ultralight structural materials and molecular sieves. In this work, a molecular modeling protocol was developed to model the dynamic polymerization of thermosetting resins into complex nanoscale architectures. The protocol relies on several recently added features of REACTER, a versatile method for modeling chemical reactions in classical molecular dynamics simulations. To demonstrate the formation of a complex phase, the polymerization of a high-temperature resin is constrained to regions of 3D space based on the isosurface of a gyroid, a triply periodic minimal surface. Morphologies of a simpler diamond lattice are also created using the same resin. Large-scale atomistic models (one million atoms) are used to generate periodic structures with lattice constants >40 nm, approaching those observed experimentally. Mechanical properties, including ultimate strength and elastic and bulk moduli, are calculated with respect to the porosity and type of lattice, and the amount of plastic deformation as a function of strain is quantified. Finally, the diffusive transport of water and hydrogen within the open channels of these nanoscale lattices is measured via molecular dynamics. The computational protocol developed is broadly applicable to obtaining atomistic insight into the mechanical and transport properties of nanoscale functional metamaterials.

525i Hybrid MD/MC Simulations of Adsorption on Flexible Nanoporous Materials

Shivam Parashar, Alexander Neimark

Nanoporous materials are known to deform upon the adsorption of guest molecules. This occurs due to the stress exerted by the guest molecules on the framework atoms resulting in the expansion or contraction of the adsorbent structure. To gain theoretical insights into this phenomenon, various simulation techniques have been developed including the osmotic ensemble method and the hybrid MD/MC simulations. The hybrid osmotic ensemble involves running short molecular dynamics simulations in the NPT ensemble as an additional Monte Carlo move in the grand canonical ensemble. On the other hand, the hybrid MD/MC requires switching between the GCMC and NPT MD simulations until the convergence is reached. In this study, we explore the differences between these two approaches using methane adsorption on IRMOF-1 as an example.

We conducted the osmotic ensemble simulations using RASPA and the hybrid MD/MC simulations using LAMMPS. Additionally, we compared the isotherms on the flexible structures with those obtained by the GCMC simulations on rigid structures and explain the observed differences, see Figure. Finally, we discuss an optimal balance between the numbers of MD and MC moves to achieve a faster convergence. Although the hybrid simulations have shown promise in flexible materials, a significant challenge that remains is the requirement for a precise flexible forcefield that can effectively reproduce the experimental isotherms.

This work is supported by the NSF CBET grant 1834339.

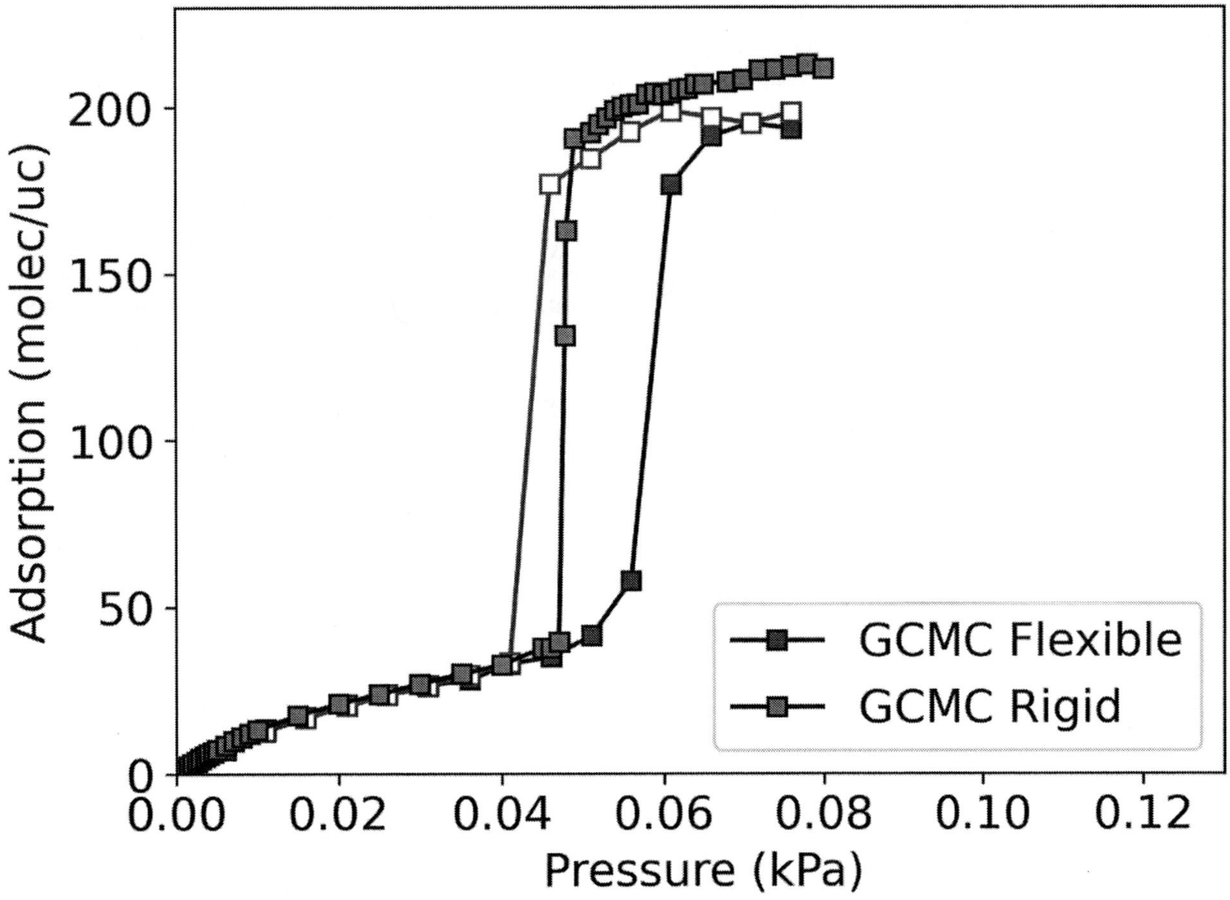

525j Prediction of Metal-Organic Framework Adsorption Isotherms Using Ab Initio Derived Neural Network Potentials

Pedro Guimaraes Martins, Yusuf Shaidu, Eric Taw, Alex Smith, Jeffrey B. Neaton

Metal-organic frameworks (MOFs) form an extensive class of porous materials of interest for new applications, as they show high surface area, high thermal stability, and strong chemical tunability. As a consequence, predicting their properties computationally is crucial in advancing the exploration of such a wide chemical space. Particularly, MOFs are promising materials in the carbon capture space, as they present low gas transport limitations, and specific MOFs have been shown to selectively and reversibly bind to carbon dioxide [2]. In this space, a predictive and generalizable method for the calculation of adsorption isotherms in MOF systems could not only accelerate the discovery of new promising materials for carbon capture but also reveal the underlying mechanism of adsorption processes.

While empirical force fields can be combined with grand canonical Monte Carlo (GCMC) simulations to estimate adsorption isotherms in specific MOFs [1], this approach relies on fixed functional forms as well as an extensive system-specific parameterization, limiting its applicability to a broader set of MOFs or to automated force field generation. Ab initio neural network potentials (NNPs), derived from density functional theory calculations, are a promising and generalizable alternative for accurate binding energetics and isotherm prediction. Furthermore, its ability to capture interactions in a diversity of chemical environments (upon further training) provides an opportunity to accurately elucidate adsorbate pore dynamics, MOF framework flexibility as well as adsorbate-adsorbate interaction effects.

In this work, we combine ab initio NNPs and GCMC simulations to predict CO2 adsorption isotherms for well-studied MOF-274, Mg2(dobpdc) (dobpdc4- = 4,4'-dioxidobiphenyl-3,3'-dicarboxylate), comparing to prior simulations and experiments. In the attached figure, we present preliminary results for this system, which shows a promising ability to recover experimental results from the proposed method. Ultimately, this method will be used to predict adsorption isotherms for different gases, such as water, nitrogen, and methane, as in application conditions additional adsorbents may be present.

In the short run, we aim at applying this prediction methodology to MOF systems which require molecular dynamics steps throughout the GCMC simulation in order to fully capture the adsorption physics. For instance, some MOFs present cooperative mechanisms related to adsorbent-adsorbent interactions, while others undergo framework mobility upon adsorption. Simulating either of these systems with parametrized force-field methods would be quite challenging, as it would be necessary to accurately capture all physical interactions in the system to correctly evolve it over time. The NNPs we propose, on the other hand, can predict the dynamics of these systems at an ab-initio accuracy, and therefore, constitute a promising method in this application space. In the long run, we hope this contribution will accelerate the understanding of gas adsorption in MOF systems, and the discovery of more efficient carbon capture materials.

[1] Mercado, Rocio, et al. 'Force field development from periodic density functional theory calculations for gas separation applications using metal–organic frameworks.' The Journal of Physical Chemistry C 120.23 (2016): 12590-12604.

[2] Sumida, Kenji, et al. 'Carbon dioxide capture in metal–organic frameworks.' Chemical reviews 112.2 (2012): 724-781.

MOF-274 CO_2 Adsorption Isotherm - 1x1x1 cell

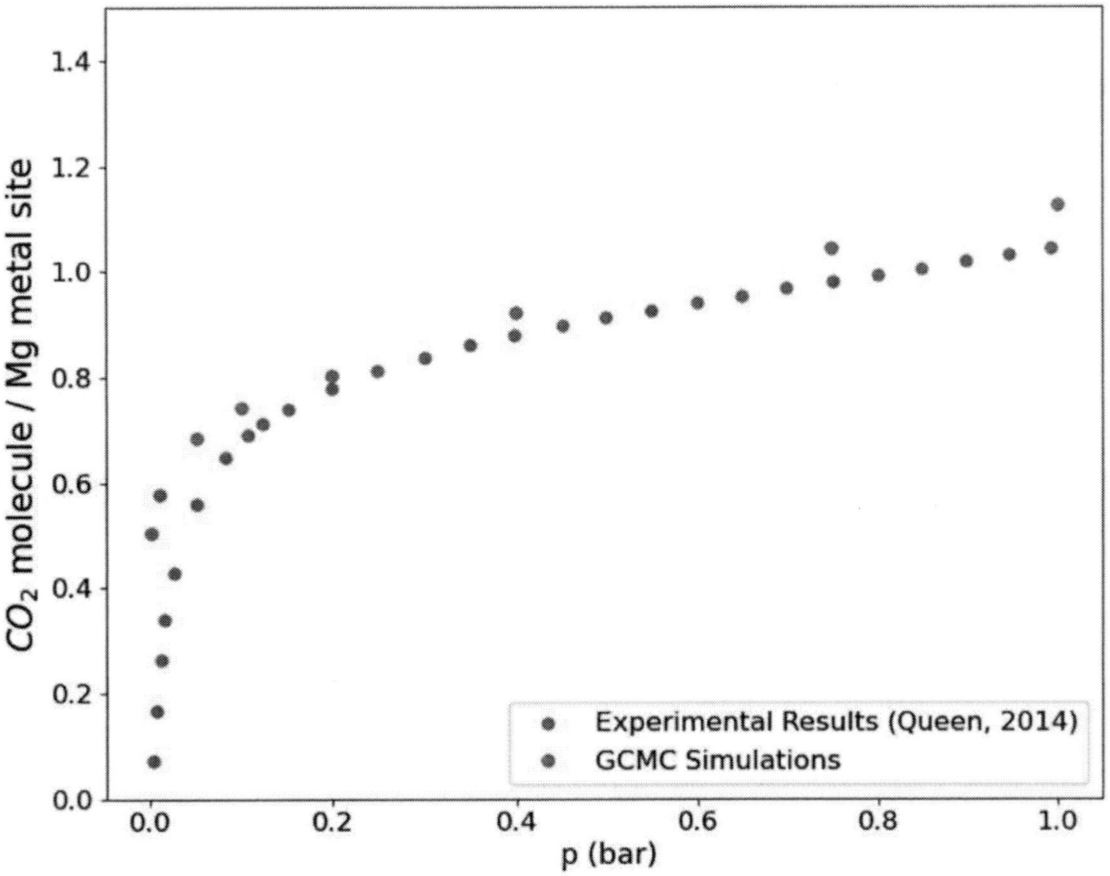

525k Exploiting Machine Learned Interatomic Model Architecture for Enhanced Transferability, Efficiency, and Robustness

Rebecca Lindsey

For decades, atomistic modelers interested in studying condensed phase reacting systems have been forced to choose between highly predictive yet computationally expensive first principles methods, and parametric "force field" approaches affording greater computational efficiency at the expense of accuracy and predictive power. But now, modelers have a third option in machine learned interatomic potentials (ML-IAP), which can yield models approximating quantum-based potential energy surfaces with computational expense that scales linearly with system size. These ML-IAPs have had a transformative effect by enabling "quantum accurate" simulations on previously inaccessible scales; however their (1) development for systems of more than a few atom types (especially those for which molecular chemistry is involved) necessitates large volumes of high expense (e.g., quantum) training data, and resulting models, and (2) they remain significantly more computationally intensive than classical "force field" models. In this presentation, we discuss recent advances in underlying form and fitting strategies for the ChIMES ML-IAP that aim to resolve these challenges by increasing model transferability and computational efficiency. Findings are discussed within the context of complex reacting materials.

525l Force Field Development for Molecular Simulations with Structure Optimized Potential Refinement

Brennon L. Shanks, Michael P. Hoepfner

One of the fundamental applications of molecular simulations is to predict the atomic structure and self-assembly of fluid-state materials. However, recent advances in experimental neutron and X-ray diffraction are revealing that many widely used force fields do not predict atomic structures that are consistent with experimental data. Directly reconstructing interatomic potentials from structure, the so-called inverse problem of statistical mechanics, is a possible method to improve and develop force fields that accurately model experimentally observed structures. While inverse techniques such as iterative Boltzmann inversion and relative entropy minimization have been successfully applied to coarse-grain model systems, there are no existing techniques that provide satisfactory solutions for real physical systems. In this study, we present a machine learning assisted structure refinement technique, called structure optimized potential refinement, that can reconstruct transferable force fields from a single neutron diffraction measurement in monatomic fluids. Applying this technique to the noble gas series (Ne, Ar, Kr, and Xe), we recover transferable force field parameters that provide excellent agreement to experimental structure and reproduce macroscopic thermodynamic properties. This result is the first demonstration that experimental scattering data contains sufficient information to predict underlying force parameters at the sub-angstrom length scale. Applications of structure optimized potential refinement include force field optimization for structure-based applications, modeling fluids in extreme environments, and predicting quantum many-body effects in fluid systems.

508a Machine Learning Enabled Development of Accurate Force Fields for Hydrofluorocarbons

Ning Wang, Montana Carlozo, Eliseo Marin Rimoldi, Alexander Dowling, Edward Maginn

Hydrofluorocarbon (HFC) refrigerants with zero ozone depleting potential have replaced chlorofluorocarbons and are now ubiquitous. However, some HFCs have high global warming potential, which has led to calls by governments to phase these HFCs out. Technologies to recycle and repurpose these HFCs need to be developed. Therefore, thermophysical properties of HFCs are needed over a wide range of conditions. Molecular simulations can help understand and predict thermophysical properties of HFCs. The prediction capability of a molecular simulation is directly tied to the accuracy of the force field. In this work, we applied and refined a machine learning based workflow to optimize the Lennard-Jones parameters of classical HFC force fields for HFC-143a (CF_3CH_3), HFC-134a (CH_2FCF_3), R-50 (CH_4), R-170 (C_2H_6), and R-14 (CF_4). Our workflow involves liquid density iterations with molecular dynamics simulations and vapor-liquid equilibrium (VLE) iterations with Gibbs ensemble Monte Carlo simulations. Support vector machine classifiers and Gaussian process surrogate models save considerable simulation time and can efficiently select optimal parameters from half a million data sets. Excellent agreement of simulated liquid density, vapor density, vapor pressure, enthalpy of vaporization, and critical properties relative to experiments were obtained. The performance of each new parameter set was superior or similar to the best force field in literature. We find that the new force fields both yield excellent VLE properties and are transferable to a broader scope of properties including transport properties. Future work will use these simulation results as the training set to calibrate a generalized force field for HFCs, and this workflow is extensible to other complex molecular systems.

508b Quantitative in silico Prediction of the Rate of Protodeboronation By a Mechanistic Density Functional Theory-Aided Algorithm

Daniel Wigh, Matthieu Tissot, Patrick Pasau, Jonathan Goodman, Alexei A. Lapkin

Boronic acids (BAs) have long been of interest to the chemical sciences community due to their irreplaceable role in communications, materials, and medicine. BAs also feature as the nucleophilic coupling partner in several key organic reactions including the Suzuki–Miyaura reaction, Chan–Evans–Lam coupling, Liebeskind–Srogl coupling, and oxidative Heck and can also undergo addition with carbonyls, imines, and enones. A reaction will not work if one of the coupling partners degrades at a significantly faster rate than the rate of the intended reaction, and protodeboronation is one of the most significant, if not the most significant, degradation pathways/side reactions in the aforementioned named organic reactions. Being able to accurately predict the rate of protodeboronation would therefore be hugely beneficial to anyone working with BAs, as it can help with reaction planning.

We have developed an algorithm for *in silico* prediction of the rate of protodeboronation of boronic acids as a function of pH [1]. Protodeboronation can proceed through 7 distinct pathways, though for any particular boronic acid only a subset of the mechanistic pathways are active [2]. The rate of each active mechanistic pathway is linearly correlated with its characteristic energy difference, which can be determined through energy calculations using Density Functional Theory (DFT). To the best of our knowledge, this is the first purpose-built algorithm to be published in the literature capable of making in silico predictions of the rate of protodeboronation. We validated the algorithm using leave-one-out cross-validation on a dataset of 50 boronic acids (469 data points from pH 0 to 14), which showed that the algorithm can explain a majority of the variability in the dataset (R2=0.65), with a Mean Absolute Error (MAE) of 0.86. The rate is measured on a log scale from -9 (22 years half-life) to +2 (7 milliseconds half-life), so MAE below 1 means that the algorithm is capable of predicting the rate of protodeboronation within an order of magnitude, i.e., whether the rate will be on the order of milliseconds, seconds, hours, days, weeks, or years. With such predictions, chemists performing reactions that feature boronic acids, such as Suzuki–Miyaura and Chan-Evans-Lam couplings, are able to better plan their reactions and predict when protodeboronation might become a problem, thus saving valuable time and materials.

Considering the abstract figure, the steps in the algorithmic protodeboronation prediction workflow can be summarised as follows:

1. DFT optimisation

The mechanistic pathway for determination of the characteristic energy difference is of the form t*ransition state -> stable intermediate*, and the energies of each of the molecular structures were determined using DFT with Gaussian (computational approach was M06L/6-311++G** in water) [3]. The DFT energy calculations for the transition states were transition state optimisations, while calculations for reaction intermediates were optimised to a local energy minimum.

2. Correlate energies

We found that there is a linear relationship between the maximum/saturated rate of a mechanism and the associated characteristic energy difference. Linear regression for the k_2 mechanism can be seen in the abstract figure. Using these linear relationships, the rate can be predicted for novel boronic acids.

3. Algorithmic amalgamation

The observed rate of reaction is the sum of the rate contributions from each active mechanism. The experimental rate measurements span from +2 (half-life of roughly 7 milliseconds) to -9 (half-life of roughly 22 years), and the predicted rate has been capped to fall within this region as well. Leave-one-out cross-validation (LOOCV) was used to assess the quality of the fit for each molecule, and the construction of a parity plot showed that the model was able to explain a majority of the variability ($R2=0.65$).

4. De novo prediction

Finally, the algorithm was used to predict *in silico* the rate of protodeboronation as a function of pH for 50 novel boronic acids of academic and industrial relevance. These predictions were made available in the supplementary information [1], which allows chemists running reactions featuring boronic acids to simply look up the predicted rate of protodeboronation for these molecules. The associated code can be found on GitHub, such that anyone can generate their own predictions. We also hope that this work may serve as inspiration as a method for predicting the rate of reaction when the mechanistic details about a reaction are known and can be simplified as a set of interacting linear equations.

[1] Wigh, D. S.; Tissot, M.; Pasau, P.; Goodman, J. M.; Lapkin, A. A. Quantitative In Silico Prediction of the Rate of Protodeboronation by a Mechanistic Density Functional Theory-Aided Algorithm. *J. Phys. Chem. A* **2023**. https://doi.org/10.1021/acs.jpca.2c08250.

[2] Cox, P. Protodeboronation. Ph.D. thesis, The University of Edinburgh, 2016.

[3] Gaussian 16, Revision C.01, Frisch, M. J.; Trucks, G. W.; Schlegel, H. B.; Scuseria, G. E.; Robb, M. A.; Cheeseman, J. R.; Scalmani, G.; Barone, V.; Petersson, G. A.; Nakatsuji, H.; Li, X.; Caricato, M.; Marenich, A. V.; Bloino, J.; Janesko, B. G.; Gomperts, R.; Mennucci, B.; Hratchian, H. P.; Ortiz, J. V.; Izmaylov, A. F.; Sonnenberg, J. L.; Williams-Young, D.; Ding, F.; Lipparini, F.; Egidi, F.; Goings, J.; Peng, B.; Petrone, A.; Henderson, T.; Ranasinghe, D.; Zakrzewski, V. G.; Gao, J.; Rega, N.; Zheng, G.; Liang, W.; Hada, M.; Ehara, M.; Toyota, K.; Fukuda, R.; Hasegawa, J.; Ishida, M.; Nakajima, T.; Honda, Y.; Kitao, O.; Nakai, H.; Vreven, T.; Throssell, K.; Montgomery, J. A., Jr.; Peralta, J. E.; Ogliaro, F.; Bearpark, M. J.; Heyd, J. J.; Brothers, E. N.; Kudin, K. N.; Staroverov, V. N.; Keith, T. A.; Kobayashi, R.; Normand, J.; Raghavachari, K.; Rendell, A. P.; Burant, J. C.; Iyengar, S. S.; Tomasi, J.; Cossi, M.; Millam, J. M.; Klene, M.; Adamo, C.; Cammi, R.; Ochterski, J. W.; Martin, R. L.; Morokuma, K.; Farkas, O.; Foresman, J. B.; Fox, D. J. Gaussian, Inc., Wallingford CT, 2016.

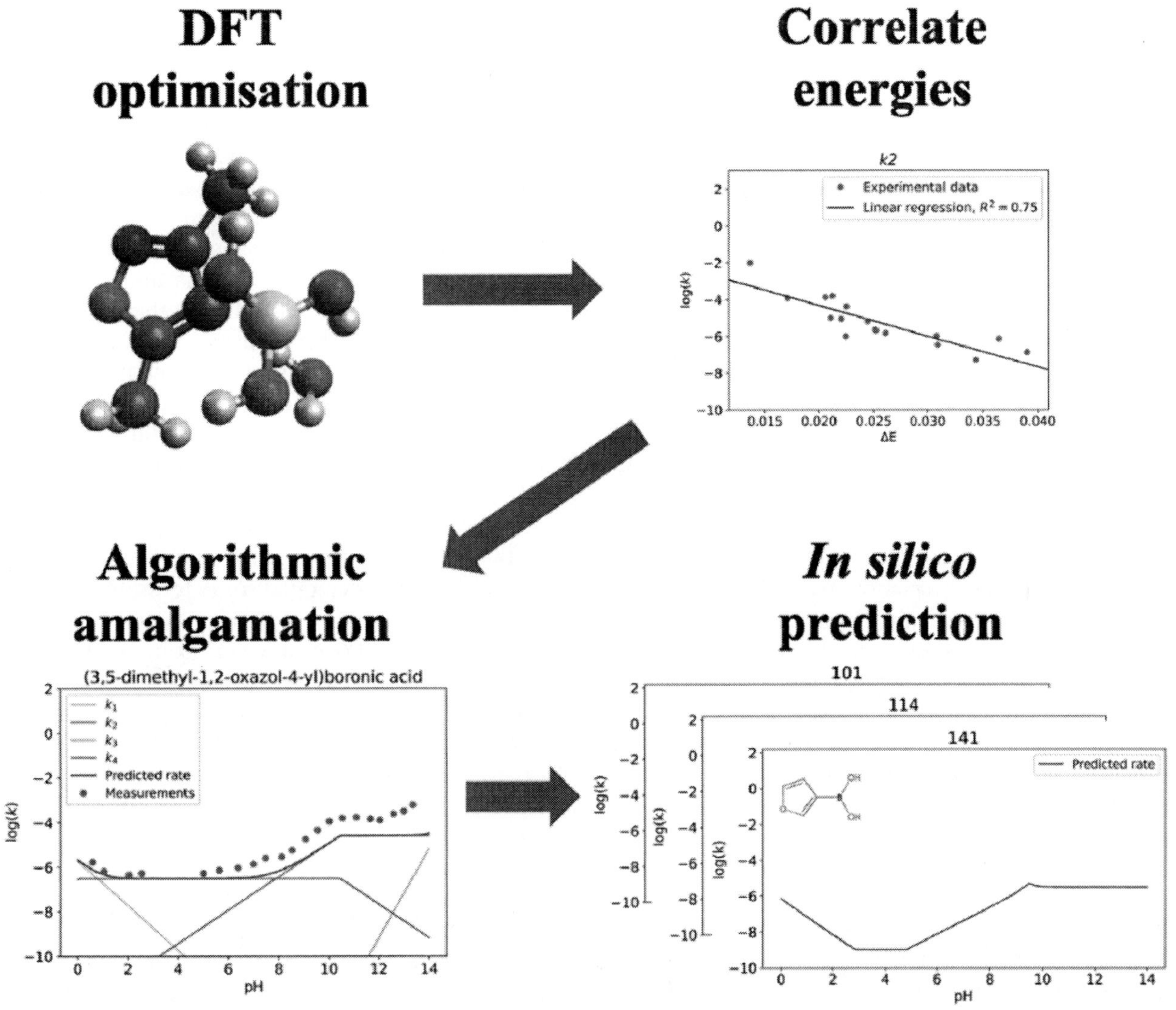
DFT optimisation
Correlate energies
k2
Experimental data
Linear regression, $R^2 = 0.75$
log(k)
ΔE
Algorithmic amalgamation
(3,5-dimethyl-1,2-oxazol-4-yl)boronic acid
k_1
k_2
k_3
k_4
Predicted rate
Measurements
pH
In silico prediction
101
114
141
Predicted rate

508c Assessment of Thermodynamic Models Using Brown's Characteristic Curves

Maximilian Urschel, Simon Stephan

For several applications, thermodynamic properties of fluids need to be modeled at high pressure and temperature, e.g. in rocket propulsion systems, geology, and astronomy. However, in parts, no experiments for determining thermophysical property data are feasible today at such conditions. Using molecular thermodynamics, physically-based models can be used for making predictions of the thermodynamic behavior at such conditions. Their reliability at such regimes is evidently – due to the lack of experimental data – a priori unknown. Hence, thermodynamic consistency tests play an important role for the assessment of the reliability of models at such conditions. An interesting strategy for testing the thermodynamic consistency is based on Brown's characteristic curves. E.H. Brown proposed four characteristic curves that define lines on the thermodynamic surface where special thermodynamic conditions hold [1]. There is one first-order curve (called Zeno curve) and three second-order characteristic curves (called Amagat, Boyle, and Charles curve), cf. Fig. 1-left. These curves are located within a large pressure and temperature range. For a given molecular fluid, Brown's characteristic curves are known to exhibit certain thermodynamic features [1,2]. The behavior of a given thermodynamic model can be tested regarding these features to assess its consistency. In this work, both molecular simulation and molecular-based equation of state (EOS) models are assessed regarding the thermodynamic consistency of Brown's criteria. In a first step, a rigorous and generalized simulation method was developed for determining the characteristic curves from a given force field [3].

The method was tested using multiple model fluids and was then applied to complex real fluids such as alkanes, alcohols, and aromatic components. For the model fluids, the classical Lennard-Jones fluid, different Mie fluids, and Stockmayer fluids were studied. For the real substance fluids, methane, ethane, propane, ethanol, and toluene (cf. Fig. 1-right) were studied. In all cases, both classical force fields and molecular-based EOS were applied. If available, also multi-parameter EOS models were used for comparison. For the model fluids, the zero-density limits of the characteristic curves were determined exactly using the virial route.

The results provide a wealth of insights in the thermodynamic consistency of such models at large pressure and temperature: The molecular force field models yield characteristic curves that are conform with Brown's criteria in all cases. This is interesting as Brown derived the criteria from rational thermodynamics for simple (repulsive-dispersive) particles only. The molecular-based EOS models considered in the evaluation show important differences, i.e. some exhibit artifacts, whereas others yield predictions that are fully conform with Brown's postulates (and show excellent agreement with the force field predictions, cf. Fig. 1-right). This is astonishing considering the fact that both the force field models and the EOS models were parametrized with respect to vapor-liquid equilibrium data alone. In some cases, also empirical EOS models were used in the evaluation that are found to exhibit artifacts in several cases. This study moreover yields insights into the applicability of the corresponding states principle at high pressure and temperature and reveals that the corresponding states principle fails at such conditions even for simple fluids. In Summary, this work contributes to the question how the quality of models can be rigorously assessed even if no experimental data is available.

References

[1] E.H. Brown: On the thermodynamic properties of fluids, Bulletin de l'Institut International du Froid Annexe 1960–1, 169–178 (1960).

[2] S. Stephan, U. Deiters: Characteristic curves of the Lennard-Jones fluid, International Journal of Thermophysics 41, 147 (2020).

[3] M. Urschel, S. Stephan: Determining Brown's characteristic curves using molecular simulation, Journal of Chemical Theory and Computation 19, 5 (2023) 1537.

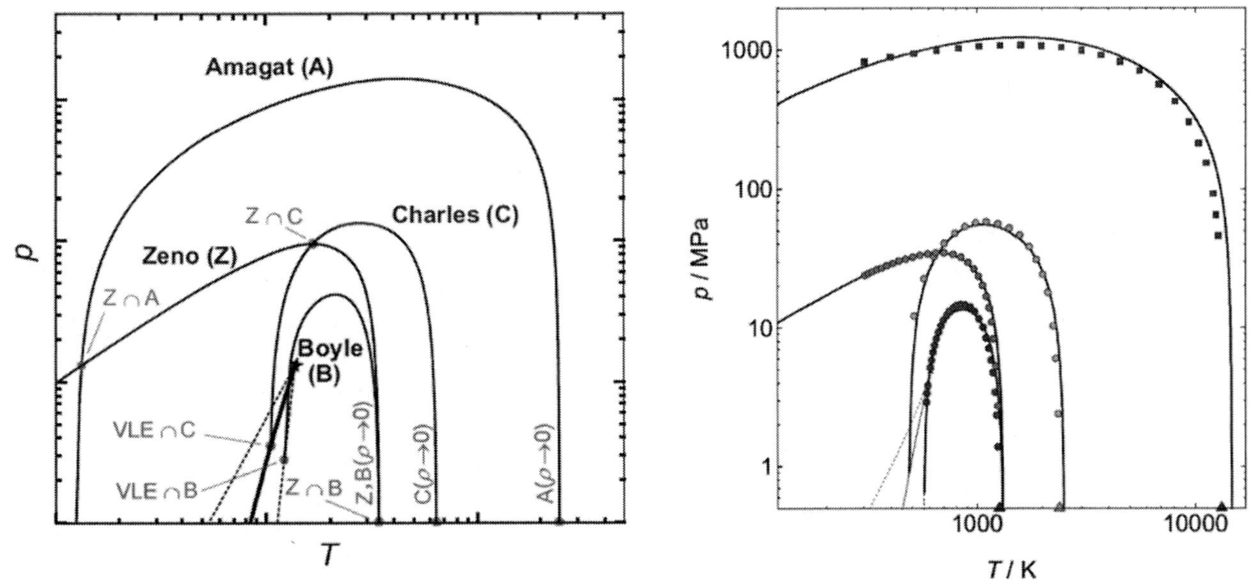

Amagat (A)
Charles (C)
Z ∩ C
Zeno (Z)
Z ∩ A
Boyle
(B)
VLE ∩ C
VLE ∩ B
Z ∩ B
Z,B(ρ→0)
C(ρ→0)
A(ρ→0)
p
T
1000
100
10
1
1000
10000
p / MPa
T / K

508d An Application of Complex Modeling in Biology: Studying Mechanochemical Dependence of Intracellular Entry Efficiency

Xinxin Wang, Zhiming Chen, Sandra Schmid, Gaudenz Danuser

Mathematical modeling of complex systems has become ever-increasingly demanded in the biology of living cells for validating experiments, generating hypotheses, and interpreting findings. Complex modeling is particularly effective for studying *in vivo* biological systems at nano-scales, where conducting experiments is challenging. Here, we built a complex model to study clathrin-mediated endocytosis (CME), the major cellular pathway of internalizing surface-bound cargo molecules. CME plays essential roles in many diseases, including COVID, cancer, and Alzheimer's. During CME, clathrin forms a self-assembled basket to consecutively bend the membrane to enclose the cargos into a ~100nm-wide vesicle within ~25 seconds before intracellular release. Multiple endocytic proteins (EP) other than clathrin also contribute to this process. However, little is clear regarding the detailed role of the EPs in assisting the clathrin basket to bend membrane mainly because of the insufficient understanding of the mechanical relation between the EPs and the cell membrane. To fill this gap, we developed a state-of-the-art stochastic model that incorporates the clathrin basket, the endocytic proteins, the cell membrane, and their mechanical interactions. We show, experimentally, that CALM (an adaptor protein) is required to initiate curvature. Next, we adjust the model based on the CALM result and the known flexibility of clathrin to predict two distinct states of the clathrin baskets at equilibrium: a shallow open state that defines abortive CME events and a curved, closed state that defines productive CME events. We further explore various values of a parameter that controls the tightness of the clathrin baskets, and discover a phase transition from the closed to open state. We hypothesize and then experimentally confirm that curvature induced by a regulatory protein dynamin2 determines the transition from open to closed states and drives productive CME. We conclude: 1) There is a checkpoint indicating a competition between the energy in clathrin baskets and the energy of membrane bending, resulting in two distinct fates of CME: abortive and productive, and 2) Dynamin2 powers clathrin baskets to pass the checkpoint and drive productive CME to enhance CME efficiency. In sum, our interdisciplinary study complementing quantitative experiments with a complex model serves as an example of bridging physics and cell biology at nanoscales.

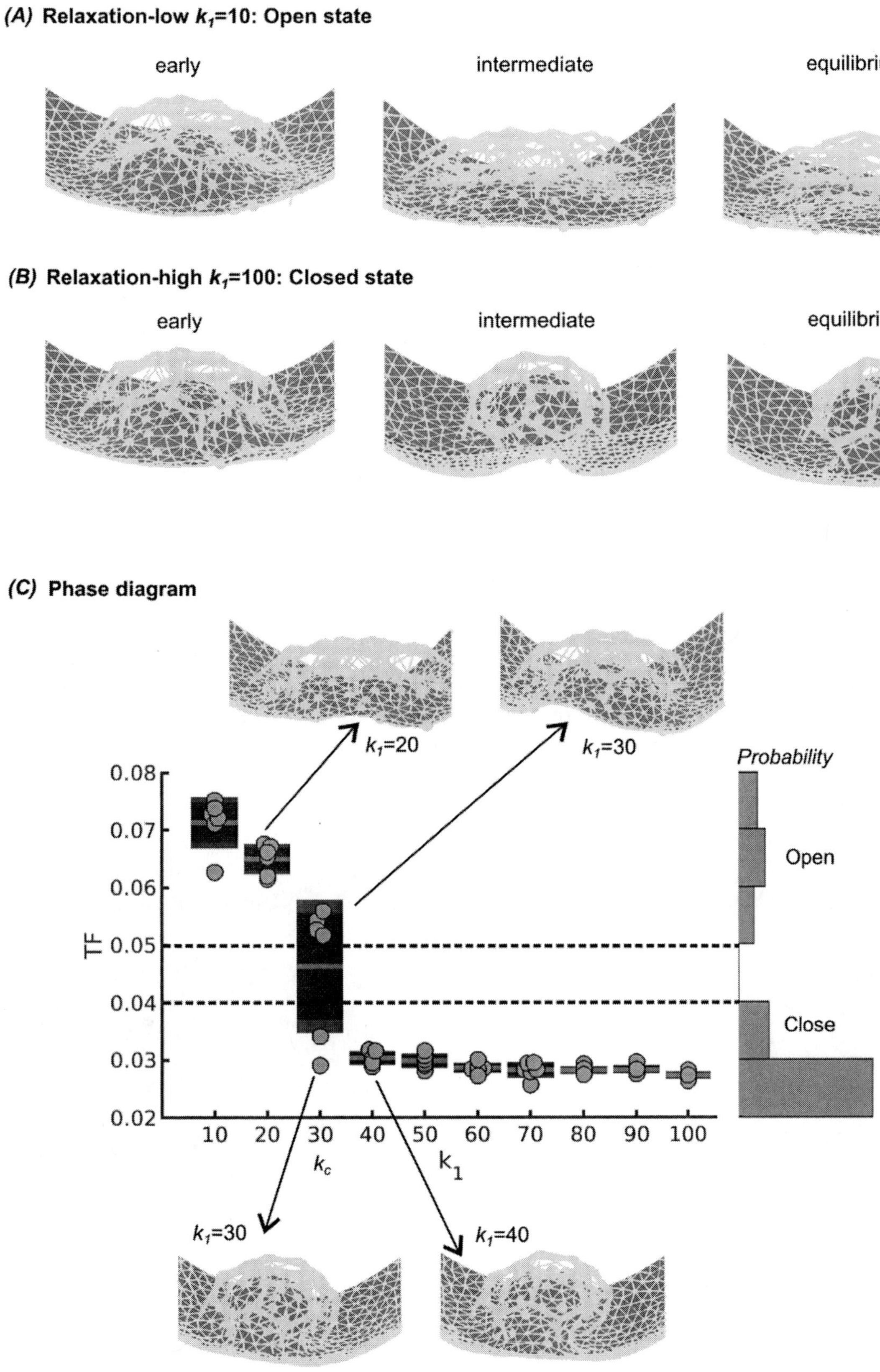
(A) Relaxation-low k_1=10: Open state
early
intermediate
equilibrium
(B) Relaxation-high k_1=100: Closed state
early
intermediate
equilibrium
(C) Phase diagram
k_1=20
k_1=30
Probability
Open
Close
TF
0.08
0.07
0.06
0.05
0.04
0.03
0.02
10
20
30
40
50
60
70
80
90
100
k_c
k_1
k_1=30
k_1=40

508f Theoretical Assessment of PET Pyrolysis Via Reactive Molecular Dynamic Simulation and Kinetic Modeling

Shuangxiu Ma, Bhavik Bakshi, Li-Chiang Lin

The rapid growth of demand for PET plastic products has resulted in a critical need to develop effective processes to reduce PET pollution by establishing a circular economy. Pyrolysis, a promising technology to produce lighter and recyclable components from wasted plastic products, has received considerable attention as a means of chemical recycling. While significant progress has been made in the experimental studying of the decomposition of PET, molecular-level observations of PET pyrolysis and its detailed reaction model remain largely missing. In this work, the pyrolysis process of PET was studied using reactive molecular dynamics (MD) simulations to elucidate its mechanism and develop kinetic models capable of describing the process. The ester oxygen-alkyl carbon bond dissociation was observed in the simulations, producing ethylene and TPA radicals. These radicals can be further transformed into long-chain components such as char, and CO_2 was released as a product. MD simulations revealed a chain reaction pattern for yielding gas species, which includes the generation of TPA radicals from long-chain breaking and condensation reactions to form a long chain with phenyl benzoate structure to consume TPA radicals. This mechanism explained the correlation between CO_2 release and long-chain formation.

Moreover, since atomistic simulations are generally conducted at the time scale of a few nanoseconds, a high temperature (i.e., > 1000 K) is typically adopted for accelerated reaction. To address the issue of the limitation of time and temperature scale, we developed a set of ordinary differential equations-based kinetic models that could describe the key products of PET pyrolysis under practical temperature scales and analyze the optimal reaction conditions. Overall, this study conducted a detailed mechanism study on PET pyrolysis and established a kinetic model with effective assessment. The workflow presented herein is capable of extracting key information from atomistic MD simulations for establishing kinetic models, which can be used in related areas for performing pre-experimental screening and mechanism prediction in order to further assist in the development of effective pyrolysis solutions. This study highlights the importance of theoretical research on pyrolysis and its potential for innovative solutions to address the growing issue of plastic pollution by developing a circular economy.

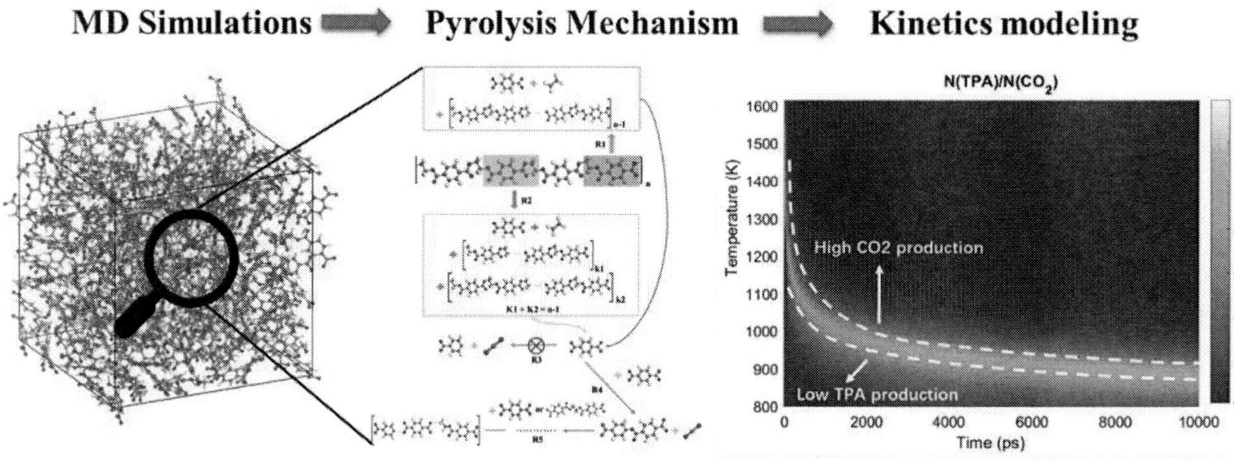

508g Efficient Simulation of Complex Fluid Phase Diagrams within Self-Consistent Field Theory Using Bayesian Optimization

Steven G. Arturo, Clyde Fare, Kaoru Aou, Huikuan Chao, Daniel Dermody, William Edsall, Jillian Emerson, Kathryn Grzesiak, Arjita Kulshreshtha, Paul M. Mwasame, Jed W. Pitera, Edward Pyzer-Knapp

The phase diagrams of complex fluids are essential for understanding solubility and miscibility, properties of central importance in polymer formulation. At a typical industrial resolution of 10 wt%, exhaustive exploration of a ternary phase diagram requires simulation or experimentation at 63 composition points, a number which grows exponentially with the number of components. Using a new objective function derived from Self-Consistent Field Theory (SCFT) computations coupled with a constrained Bayesian optimization algorithm, we demonstrate efficient search of phase boundaries in a sample two-phase ternary modeled by polymer SCFT with 50% fewer simulations than an exhaustive search. Our approach is general, gradient-free, and can be applied to either simulation or experimental campaigns.

508h Investigating the Effect of Topology on the Shape Memory Properties of Thermoset Shape Memory Polymers Using Coarse-Grained Molecular Dynamics Simulations

Pouria Nourian, Collin D. Wick, Andrew J. Peters

Thermoset shape memory polymers (TSMPs) are materials that can change shape when exposed to heat, making them useful for various applications in fields like biomedicine, aerospace, automotive, robotics, and smart textiles. This study focuses on how manipulating the topology of TSMPs can improve their properties and potential for use in additive manufacturing. Via a coarse-grained model, we investigated various ways to quantify network quality (i.e., degree of topological perfection), and demonstrated that network quality has important effects on rubbery elastic modulus, recovery stress, and recovery ratio. We also found that our best parameter to describe network quality (the topological score) is a better predictor of network properties than crosslinking fraction. Additionally, higher hardener functionality also improved material properties, especially when the topological score is also high. These results demonstrate that topology significantly affects the mechanical and shape memory properties of TSMPs, and such an effect can be quantitatively characterized in terms of network perfection and quantified by a simple parameter. This study suggests that increasing crosslinking fraction and hardener functionality, combined with methods to improve network quality, can lead to significant improvements in the shape memory properties of TSMPs.

566a *Scikit-Matter*: An Open-Source Suite of Generalizable Machine-Learning Methods Born out of Chemistry and Materials Science

Rose Cersonsky

Easy-to-use libraries such as *scikit-learn* have accelerated the adoption and application of machine learning (ML) workflows and data-driven methods. While many of the algorithms implemented in these libraries originated in specific scientific fields, they have gained in popularity in part because of their generalizability across multiple domains. Over the past two decades, researchers in the chemical and materials science community have put forward general-purpose machine learning methods. The deployment of these methods into workflows of other domains, however, is often burdensome due to the entanglement with domain-specific functionalities. Here, I present the python library *scikit-matter* that targets domain-agnostic implementations of methods developed in the computational chemical and materials science community, following the *scikit-learn* API and coding guidelines to promote usability and interoperability with existing workflows. I will highlight both uses for these methods and software within the chemical sciences and beyond.

566b Pysages: Flexible, Efficient, GPU-Accelerated Sampling Methods

Pablo Zubieta, Ludwig Schneider, Gustavo Perez Lemus, Juan J. de Pablo

Molecular Dynamics simulations are critical for understanding and predicting molecular properties, but determining the free energy of complex systems requires advanced sampling techniques. PySAGES is a Python open-source library that provides enhanced sampling methods for molecular dynamics simulations, including adaptive forces, harmonic bias, and forward flux sampling. PySAGES is designed for massive parallel applications and provides full GPU support, making it an efficient and scalable tool for exploring complex free energy landscapes. The software features an intuitive interface that facilitates inclusion of new collective variables and sampling methods. This work introduces the core features of PySAGES and provides concise examples of its capabilities, including benchmark results that demonstrate its performance. PySAGES is a versatile and efficient tool for studying complex molecular systems, advancing our understanding of molecular properties, and enabling the design of new materials.

*This work was supported by MICCoM, as part of the Computational Materials Sciences Program funded by the U.S. Department of Energy, Office of Science, Basic Energy Sciences, Materials Sciences, and Engineering Division through Argonne National Laboratory.

566c The Molecular Simulation Design Framework (MoSDeF): The Latest Integration with Gomc, Cassandra, and HOOMD-Blue

Co D. Quach, Nicholas C. Craven, Brad Crawford, Ryan S. DeFever, Jenny Fothergill, Christopher Iacovella, Jeffrey Potoff, Edward Maginn, Eric Jankowski, Clare McCabe, Peter Cummings

The Molecular Simulation Design Framework (MoSDeF)[1] was developed as a suite of tools to alleviate issues around initializing and performing molecular simulations with an emphasis on facilitating reproducibility in molecular simulation.[2] By providing standardized, open-source libraries, the MoSDeF can automate steps in the preparation of chemical/biological systems for simulation, and minimize unnecessary human interaction and associated errors. Furthermore, automation of workflows allows for easier setup of high-throughput screening processes, by integrating with workflow management tools such as signac and signac-flow.[3] MoSDeF has been successfully applied to facilitate simulation studies in conjunction with several open source simulation codes. From the beginning, the MoSDeF project has provided support for molecular dynamics simulation engines, such as GROMACS and LAMMPS.[4, 5] As part of the continued community development of MoSDeF support has now been extended to GOMC (via the MoSDeF-GOMC library), Cassandra (via the MoSDeF-Cassandra library), and HOOMD-blue.[6-8] These integrations enable more automated, intuitive, and reproducible ways to set up simulations with these engines. MoSDeF can be used to easily build individual systems for new users and complex/integrated reproducible workflows for experts. This enables these workflows to be used in teaching new users and running training workshops without the usual overhead associated with learning inputs to complex codes. We will continue working to further integrate MoSDeF with other tools/simulation engines existing in the community, joining forces with similar efforts, such as the FAIR consortium and OpenKIM, to effectively address the reproducibility issue of the field.[9, 10]

References

1. “MoSDeF” [Online]. Available: https://mosdef.org
2. Cummings, P. T., MCabe, C., Iacovella, et al. (2021). Open-source molecular modeling software in chemical engineering focusing on the Molecular Simulation Design Framework. In AIChE Journal (Vol. 67, Issue 3). Wiley. https://doi.org/10.1002/aic.17206
3. “Signac” [Online]. Available: https://signac.io
4. Abraham, M. J., Murtola, T., Schulz, R., et al. GROMACS: High Performance Molecular Simulations through Multi-Level Parallelism from Laptops to Supercomputers. SoftwareX, 2015, 1–2, 19–25. https://doi.org/10.1016/j.softx.2015.06.001.
5. Thompson, A. P., Aktulga, H. M., Berger, R. et al. LAMMPS - a Flexible Simulation Tool for Particle-Based Materials Modeling at the Atomic, Meso, and Continuum Scales. Computer Physics Communications, 2022, 271, 108171. https://doi.org/10.1016/j.cpc.2021.108171.
6. Crawford, B., Timalsina, U., Quach, C. D., et al. MoSDeF-GOMC: Python Software for the Creation of Scientific Workflows for the Monte Carlo Simulation Engine GOMC. Journal of Chemical Information and Modeling, 2023, 63, 1218–1228. https://doi.org/10.1021/acs.jcim.2c01498.
7. DeFever, R. S., Matsumoto, R. A., Dowling, A. W., et al. MoSDeF Cassandra: A Complete Python Interface for the Cassandra Monte Carlo Software. Journal of Computational Chemistry, 2021, 42, 1321–1331. https://doi.org/10.1002/jcc.26544.
8. Anderson, J. A., Glaser, J., Glotzer, S. C. HOOMD-Blue: A Python Package for High-Performance Molecular Dynamics and Hard Particle Monte Carlo Simulations. Computational Materials Science, 2020, 173, 109363. https://doi.org/10.1016/j.commatsci.2019.109363.

9. Wilkinson, M., Dumontier, M., Aalbersberg, I. et al. The FAIR Guiding Principles for scientific data management and stewardship. Sci Data 3, 160018 (2016). https://doi.org/10.1038/sdata.2016.18
10. Tadmor, E. B., Elliott, R. S., Sethna, J. P., Miller, R. E., Becker, C. A. The Potential of Atomistic Simulations and the Knowledgebase of Interatomic Models. JOM, 2011, 63, 17–17. https://doi.org/10.1007/s11837-011-0102-6.

566d Lessons Learned and Best Practices for Interactive Workshops from the Institute for Computational Molecular Science Education

Neeraj Rai, Michael Shirts, Sapna Sarupria, Jindal Shah, Eric Jankowski

The Institute for Computational Molecular Science Education supports the development of molecular simulation pedagogy, open source curricula, and professional networks through in-person and online workshops throughout the year. Here we summarize activities across four in-person summer schools and one online school. We describe the software development techniques, strategies, and tools we apply to curriculum development that includes tutorial contributions across a variety of computational fields. We show examples of open materials that can be used by the broad molecular sciences community and discuss areas for improvement. Further we describe efforts to train workshop trainers, the experiences of learners across these five workshops. We highlight high-impact training practices, future plans and activities, and invite broad participation from those involved in chemical education, software engineering, simulation, open educational materials, and curriculum development.

This material is based upon work supported by the National Science Foundation under Grants No. 2118217, 2118204, 2118174, 2200907, and 2118180.

566e Writing Bug-Free Software Using Lean

Tyler R. Josephson

When developing new methods for molecular simulation, eliminating bugs can be challenging. Programming languages like Python, FORTRAN, and Julia flag syntax errors, but don't (and cannot) catch errors in math or program logic - these must be rooted out by human experts. These issues can be managed by following best practices in software development (such as writing tests alongside the program), but even these do not guarantee that code is correct.

Lean is a new programming language developed for writing and automatically checking the logic of math proofs. We show how Lean can be used to write programs in scientific computing, and report on our progress toward writing a molecular dynamics simulation engine, and proving its correctness, in Lean.

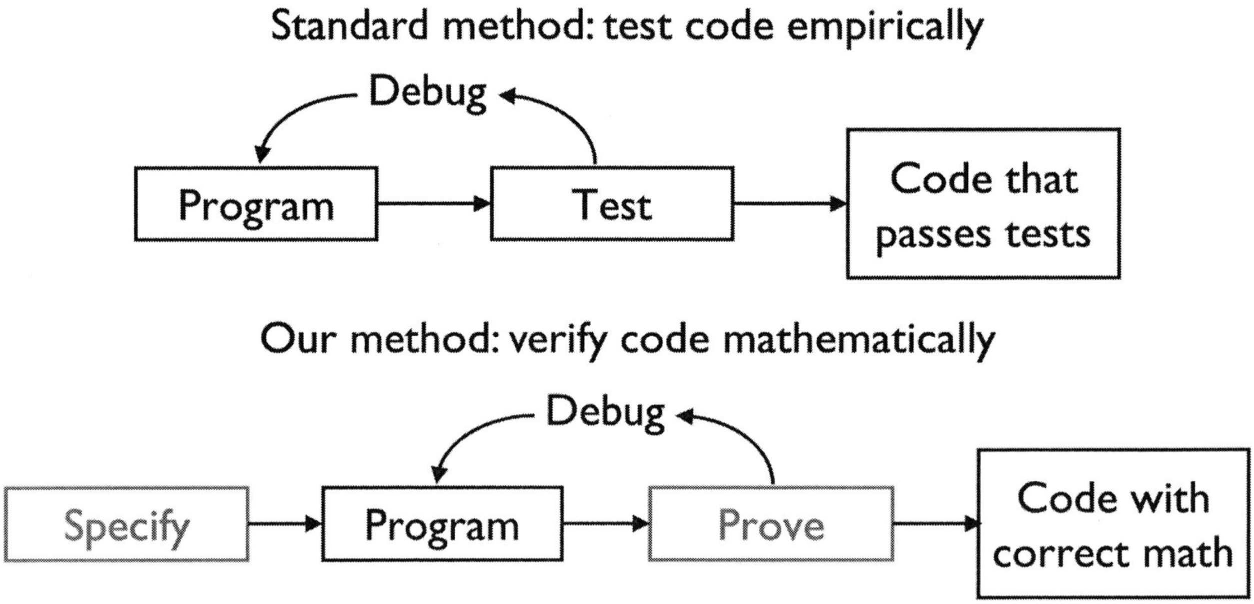

566f Ontospecies: A Dynamic Knowledge Graph for the Representation of Chemical Species.

Laura Pascazio, Simon Rihm, Jethro Akroyd, Sebastian Mosbach, Markus Kraft, Jiaru Bai

Data is beginning to play a greater role in transforming the chemical engineering research and practice. The size of chemistry databases, their complexity, and their number have increased continuously over time. One of the most comprehensive general public database is PubChem. It hosts information on more than 60 million unique chemical structures and it serves as a key chemical information resource for researchers in many biomedical science areas, including cheminformatics, chemical biology, and medicinal chemistry. PubChem provides various tools that fulfil criteria for simple and effective searching. However, these tools are insufficient if the search needs to be designed to fulfil complex criteria as well as if new information needs to be derived from existing data.

Since the landmark publication by Berners-Lee et al., the semantic web field has envisioned the next generation of the web in both a human- and machine-readable format and in recent years it is emerging as an increasingly important approach for better scientific data sharing and faster data processing using computers.

The World Avatar (TWA) project uses the semantic web technologies to create a digital 'avatar' of the real world. The digital world is composed of a dynamic knowledge graph that contains concepts and data that describe the world, and an ecosystem of autonomous computational agents that simulate the behaviour of the world and that continuously update the concepts and data. A knowledge graph (KG) is a network of data expressed as a directed graph, where the nodes of the graph are concepts or their instances (data items) and the edges of the graph are links between related concepts or instances. This provides a powerful means to host, query and traverse data, and to find and retrieve related information. The autonomous computational agents are the key aspect of the dynamic nature of the KG. They continuously and independently act on the KG performing various tasks with the aim of producing a self-growing, self-updating, and self-improving ecosystem.

To create a digital representation of the world that bridges the molecular-scale chemistry level to real world macroscale phenomena and enables cross-domain applications, it is crucial to have a rich general chemistry domain. Exposing the PubChem data to semantic web services may help in this regard. Due to the difficulties on dealing with data from different sources and mostly collected in the form of strings, the current databases that translate data from PubChem in relational databases do not include all the available information that can be accessed in the web. Properties like boiling point, melting point, density or solubility as well as spectral information on chemical species are currently not available in any relational database.

The purpose of this work was to develop an ontology, OntoSpecies, that describes chemical species and their properties and that aims to serve as core of the chemistry domain of TWA KG and at the same time address some of the limitations of previous chemistry relational databases. Specifically, the resulting ontology:

- Provides knowledge on general chemistry concepts related to chemical species through the integration of PubChem data on compounds into the database using a software agents. Concepts that are currently not exported in any relational database are also included in the ontology.

- Gives access to the dataset through a SPARQL endpoint. This will remove the inherent limitations of using the web-based PubChem resource (such as inability to construct complicated queries using the available web-based interfaces) by allowing a researcher to use readily available semantic technologies to query and analyze PubChem data on local computing resources.

- Uses a dynamic knowledge-graph-based approach to have a self-growing database that not only integrates data from different sources but also creates and infers knowledge through the uses of software agents.

It is anticipated that our approach will play a key role in the next generation of chemical informatics. The ontological format permits advanced queries, and easy data analysis and visualization. This can be used to compare chemical properties of similar compounds, find compounds with required characteristics as well as automate laborious data gathering from researcher. We show how tasks like the identification of species in unknown mixture based on NMR spectrum, the selection of suitable solvents based on multiple criteria or the investigation of trends in chemical properties can be addressed using SPARQL queries in combination with the use of software agents to postprocess the information obtained as query result. We also show how the ontological format is beneficial to maintain and enrich the data set, as well as to check the consistency and accuracy of the data. Finally, the link between OntoSpecies and other ontologies in TWA is discussed in the context of laboratory automation and cross-domain applications in TWA ecosystem.

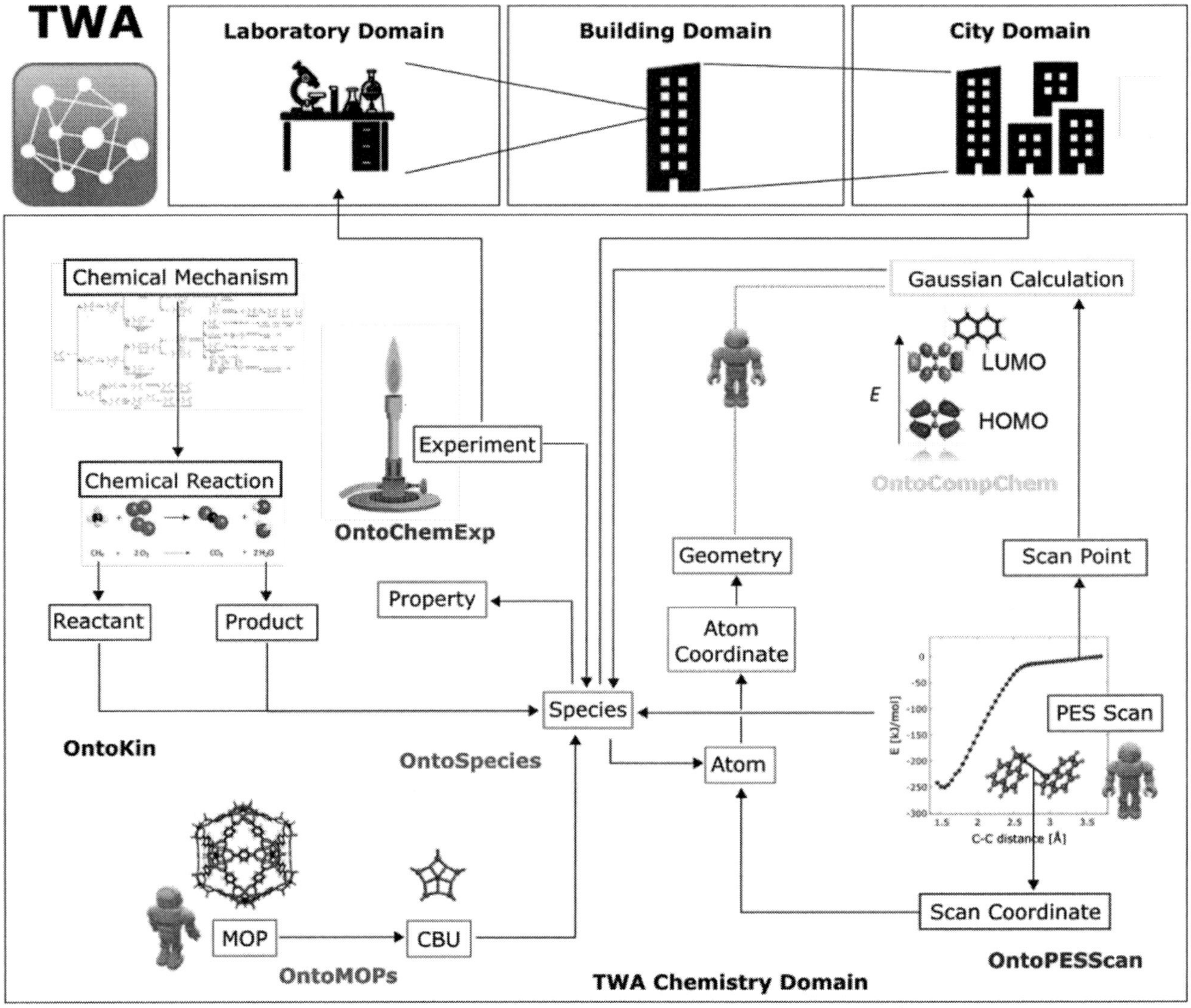

566g Molmod: A Force Field Database for Molecular Simulations of Fluids

Simon Stephan, Sebastian Schmitt, Florian Fleckenstein, Hans Hasse

The MolMod force field database [1] is presented. The MolMod database is an open-access web-based database for classical force fields for fluids that can be used in molecular dynamics or Monte Carlo simulations for modeling a large variety of thermodynamic properties as well as for the simulation of nanoscopic processes. The MolMod database contains both component-specific force fields and transferable force fields. In a component-specific force field, the model parameters of a given force field are only valid for a certain component. A transferable force field, on the other hand, is a chemical construction plan specifying intermolecular and intramolecular interactions between different types of atoms or different chemical groups and can be used for building a force field model for a given component. The vast majority of force fields in the database was parametrized using vapor-liquid equilibrium data and accordingly describe such data well. In many cases, also predictions of other properties, like transport and interfacial properties, were tested and found to be in good agreement with experimental data. Many of the molecular models were also successfully tested for modeling mixtures using combination rules like those of Lorentz and Berthelot.

The MolMod database presently comprises approximately 700 component-specific force fields – and is constantly growing. This includes mostly molecular substances and also some ionic components. The database comprises both rigid force fields as well as flexible force fields with internal degrees of freedom. Moreover, the MolMod database comprises eight transferable force field that can be used for describing a large number of components. Force field developers are invited so submit their models to the MolMod admins for being included in the database.

For transferable force fields, the MolMod database contains a graphical user interface in the front-end of the website that enables the user to construct essentially any organic molecule. Moreover, the user can specify, which transferable force field is to be used, e.g. TraPPE or OPLS. Then, the website provides the actual force field parameters for the chosen molecule (cf. Fig. 1).

For implementing transferable force fields in the MolMod database, a new SQL-based data scheme was developed, which provides a machine-readable format for integrating a given transferable force field. This force field scheme is generalized such that different transferable force fields can be formulated within the data scheme. The data scheme can be applied to all-atom as well as united-atom transferable force fields. Using this data scheme, multiple popular transferable force fields are implemented in the database, e.g. the TraPPE, OPLS-AA, and Potoff force field.

The MolMod database provides a large variety of search functionalities. For all force fields, the references to the original literature are given. The force fields are presented in a consistent unit system containing the full specifications of the molecular model. Moreover, the MolMod database provides ready-to-use input files for different simulation engines, e.g. LAMMPS [2], GROMACS [3], ms2 [4], and ls1 mardyn [5] etc. Since these molecular simulation codes are based on different approaches to realize for example multipole interactions, the geometry definition, and the rigidity of molecules, on the fly conversion routines are implemented in the database kernel. This enables researchers to include the molecular models easily in their individual work flows for predicting thermodynamic properties of fluids.

References:

[1] S. Stephan et al.: Mol. Simul. 45 (2019) 806.

[2] S. Plimpton et al.: J. Comp. Phys. 117 (1995) 1.

[3] M.J. Abraham et al.: SoftwareX 1 (2015) 19.

[4] C.W. Glass et al.: Comp. Phys. Comm. 185 (2014) 3302.

[5] C. Niethammer et al.: J. Chem. Theor. Comp. 10 (2014) 4455.

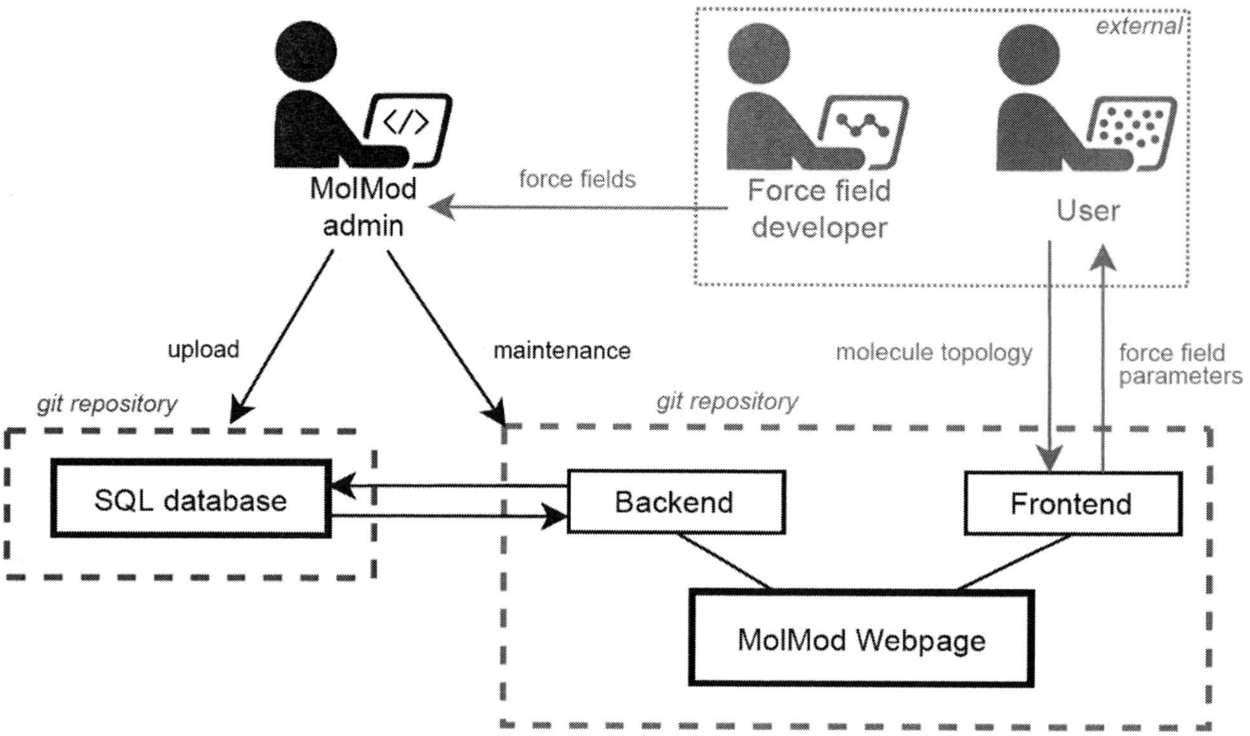
external
MolMod
admin
force fields
Force field
developer
User
upload
maintenance
molecule topology
force field
parameters
git repository
git repository
SQL database
Backend
Frontend
MolMod Webpage

566l Type Labels in LAMMPS: Enabling New Capabilities for OpenKIM and REACTER

Jacob Gissinger

Type labels are alphanumeric strings assigned to atoms, bonds, angles, etc. to supplement the numeric types that are traditionally used in LAMMPS. The introduction of type labels into LAMMPS helps to streamline several types of common tasks and workflows, and opens up many possibilities for new features. For example, atom types could be assigned labels that match the conventions of a given force field, which means LAMMPS atom types are human readable and compatible with tools that follow the same conventions. Input files are more universal when they use type labels rather than assign arbitrary numbers to interaction types for each simulation. The increased portability of topology files greatly simplifies the task of setting up simulations that require multiple data files or molecule template files. Workflows involving LAMMPS features that require creating multiple files or manipulating interaction types, such as the REACTER protocol, are considerably streamlined once adapted to take advantage of type labels. Type labels enable advanced capabilities within the REACTER protocol, including the automatic assignment of types for new bonded interactions created during reactions. The type labels framework also allows bonded force fields to be supported within the OpenKIM repository. The developers look forward to working with the community so that more molecular modeling tools can take advantage of the type labels framework.

566i Pysages: Funnel Restraints for Ligand-Receptor Enhanced Sampling Calculations

Gustavo Perez Lemus, Cintia Menéndez, Pablo Zubieta, Juan J. de Pablo

The Covid-19 pandemic has underscored the pressing need for new, efficient, and accurate methods that can accelerate drug design and discovery. Molecular dynamics (MD) simulations coupled with enhanced sampling methods provide a detailed understanding of Protein-Ligand (P-L) interactions. In the past, complex restraint potentials have been coupled with the metadynamics enhanced sampling method to create Funnel Metadynamics, which calculates free energies in protein complexes. In this work, we propose the use of funnel-like potentials in combination with an expanded set of enhanced sampling methods beyond the metadynamics family to achieve faster and more efficient sampling of P-L interactions. This approach has the potential to accelerate computer-aided drug design and discovery.

566j Exploring the Reproducibility of Molecular Simulations Using the Molecular Simulation Design Framework (MoSDeF)

Nicholas C. Craven, Ramanish Singh, Co D. Quach, Chris R. Iacovella, J. Ilja Siepmann, Clare McCabe, Peter T. Cummings

Molecular simulations are increasingly used to accurately predict the properties of novel materials and understand molecular based phenomena beyond the level of current imaging techniques. Because of this, a wide array of techniques, methods, and software has been published and distributed to the scientific community. Specifically, several molecular dynamics (MD) and Monte Carlo (MC) molecular simulation codes have been made open-source and community developed in order to reduce the workload, overhead, and potential errors with in-house computational codes. Recently, Schappals et al. published work on the comparison of results for density calculations from different force fields with varying groups performing the same calculations using different simulation engines[1]. They found reasonable reproducibility, with a standard deviation of the relative errors (STD-RE) of 0.95%, but the results did not fall within statistical agreement across the engines they studied. They concluded that these systematic errors are inherent to molecular simulations once a certain degree of complexity is reached and stem from a combination of code implementation, software, and human errors. d is a software package designed to standardize inputs to different simulation software, and has the benefits of implementing models in a standardized, well tested, and scriptable way, following the principles of TRUE simulations[2,3]. This work shows that using MoSDeF to initialize a workflow results in consistent engine differences (STD-RE 0.099%) for even more complex arrays of models and methods. Comparing the bias as synonymously as possible to the methods employed by Schappals et. al., MoSDeF gives a 9.7-fold decrease in relative error on a per simulation basis, and a 43-fold decrease in error using best practices replicate averaging. Additionally, the reduction in systematic errors in the model implementation allows us to explore the contributions of other errors from more subtle sources, such as the effect of flexible vs rigid bonds (0.2%-0.8%), the effect of different long range correction treatment (1.2-3.5%), and the effect of slightly altered SPC/E forcefield parameters (0.05%).

References

1. Schappals, M.; Mecklenfeld, A.; Kröger, L.; Botan, V.; Köster, A.; Stephan, S.; García, E. J.; Rutkai, G.; Raabe, G.; Klein, P.; Leonhard, K.; Glass, C. W.; Lenhard, J.; Vrabec, J.; Hasse, H. Round Robin Study: Molecular Simulation of Thermodynamic Properties from Models with Internal Degrees of Freedom. *J. Chem. Theory Comput.* 2017, 13, 4270–4280, PMID: 28738147.
2. Summers, A. Z.; Gilmer, J. B.; Iacovella, C. R.; Cummings, P. T.; MCabe, C. MoSDeF, a Python Framework Enabling Large-Scale Computational Screening of Soft Matter: Application to Chemistry-Property Relationships in Lubricating Monolayer Films. *J. Chem. Theory Comput.* 2020, 16, 1779–1793.
3. Thompson, M. W.; Gilmer, J. B.; Matsumoto, R. A.; Quach, C. D.; Shamaprasad, P.; Yang, A. H.; Iacovella, C. R.; McCabe, C.; Cummings, P. T. Towards molecular simulations that are transparent, reproducible, usable by others, and extensible (TRUE). *Mol. Phys.* 2020, 118, e1742938.

566k Automated Molecular Structure Deduction from Spectral and Reaction Data

Tianfan Jin, Brett Savoie

Molecular structure identification is a fundamental task in chemistry that enables various niche applications such as degradant analysis, property prediction and formulation development. Spectroscopic methods, including mass spectrometry (MS), infrared (IR), and nuclear magnetic resonance (NMR) spectroscopy, are widely used to identify unknown products. However, due to the complexity of potential molecular structure space, the information obtained from spectroscopic characterization is often deficient, leaving the problem of elucidating chemical structure from spectral results underdetermined. To address this challenge, we propose a machine learning (ML) model that integrates spectral outcomes with information on starting materials, and establishes a deductive workflow for structure prediction. Our approach leverages the fact that researchers usually have information on what they put in the flasks prior to analyzing the unknown spectroscopic data. This additional information of reactants and reagents restricts the vastness of product space, leading to more precise structure prediction over analyzing spectra from scratch. The deductive ML model comprises several parallel-placed transformer modules which study the structural features of reactants, as well as the spectral fingerprint patterns. A deductive super-network then comprehensively deduces the most likely molecular candidates. It is typical that the success of ML approach depends on the quality and completeness of the data used to train the model. However, our model has the advantage of maintaining comparable performance even when one or more spectroscopes are absent.

107a What Can Molecular Modeling and Machine Learning Teach Us about Ionic Liquid-Ionic Liquid Mixtures?

Jindal Shah

Ionic liquids are a class of non-aqueous solvents comprised entirely of ions. The asymmetry in structural motif of the ions and charge delocalization result into a large number of ionic liquids with melting points below room temperature. Over the last two decades, ionic liquids have drawn considerable attention from the research community due to the fact that many ionic liquids are practically non-volatile, thermally, chemically, and electrochemically stable. Additionally, a large number of cations and anions are available so that an ionic liquid with properties desirable for a given application or a process can be designed, at least in theory. A major theme of these investigation is on elucidating structure-property relationships of pure ionic liquids by a careful selection of the cation, anion, and/or substituents on the ions. At molecular-level, the extent to which such changes affect a delicate balance of electrostatic, hydrogen bonding, and van der Waals interactions will determine how different the macroscopic properties will be from those of the parent ionic liquid.

Our effort, over the past several years, has focused on understanding a connection between changes in the molecular-level interactions that occur in ionic liquid-ionic liquid mixtures and their impact on microscopic and macroscopic properties. In this presentation, we will discuss several ionic liquid mixtures in which the difference in the molar volume and hydrogen bonding ability of anions drive local structural reorganization of ions and long-range structural order in ionic liquid-ionic liquid mixtures, which is absent in pure ionic liquids. We will demonstrate that these structural transitions lead to varying gas dissolution mechanisms and a maximum in dielectric constant. Furthermore, we will showcase how machine learning can be leveraged to provide clues on which ionic liquid mixtures are likely to exhibit a maximum or a minimum in property such as ionic conductivity warranting a detailed molecular simulation-based investigation.

107b Navigating the Landscape of Advanced Materials

Yamil Colón

The fourth industrial revolution will be facilitated by the discovery of advanced materials that increase the interconnection of technology and human life. One of the main challenges is identifying promising material candidates for a target application given the virtually limitless number of material possibilities. Computer simulations are an integral tool to survey the material landscape and their performance. In this talk, I will present some of our efforts in the context of metal-organic frameworks (MOFs) which are nanoporous, crystalline materials that combine inorganic and organic building blocks. I will first present on machine learning (ML) efforts to characterize the material landscape and their performance in multiple adsorption applications. I will highlight transfer learning and active learning algorithms to make new predictions at dramatically reduced computational cost, helping navigate the material and thermodynamic conditions required for material performance evaluation. Lastly, I will touch upon the non-traditional application of the developed computational toolset for MOF discovery in quantum optics applications.

107c Navigating and Sculpting Structure-Function Landscapes of Polymeric Materials

Michael Webb

The chemical design of polymers with target structural and/or functional properties represents a grand challenge in materials science. Although artificial intelligence and machine learning have greatly enhanced design efforts for many materials classes, applications to polymer systems remains limited, particularly those aiming to exploit property variation due to composition, sequence, or topological effects. In this talk, I will describe our efforts to combine molecular simulation, machine learning, and, in some cases, experimental data to navigate complex polymer design spaces in search of specific properties. I will specifically share examples related to the structure/mechanics of single-chain nanoparticles, composition-dependent phase behavior of polymer solutions, and the interactivity of polymer chains with biological interfaces. These vignettes will highlight both methodological advancements as well as intriguing applications for machine learning and simulation of soft materials.

107d Effect of Temperature and Shear on Protein Denaturation: Insights from Molecular Dynamics Simulations

Janani Sampath

Biologic drugs, or "biologics" are one of the fastest growing classes of FDA-approved therapeutics. These compounds are often fragile, and require conjugation to polymers for stabilization, with many proteins too ephemeral for therapeutic use. It is important that protein structure be preserved during conjugation and after delivery, as this dictates the efficacy of the drug. During storage and administration, proteins may unravel and lose their secondary structure due to various factors, such as fluctuations in temperature, solution pH, shearing forces, and other external stressors. Understanding the relationship between these stressors and protein structure is critical in developing effective strategies to prevent protein misfolding and aggregation. Despite the widespread use of PEG in therapeutics, the specific mechanisms by which polymers stabilize proteins remain largely unknown. A comprehensive understanding of the destabilizing effects triggered by external stressors and the subsequent stabilizing interactions of polymers is essential for designing more stable and effective protein therapeutics.

Protein-polymer conjugates typically lead to very low yields, and trial and error experimentation to determine structure-function relationships laborious, time intensive, and expensive. Computational simulations can overcome many of these limitations and connect protein structure to polymer chemistry over broad spatiotemporal scales. In this study, we employ all-atom molecular dynamics (MD) simulations to examine the unfolding process of lysozyme and insulin under high temperature and shear. By accessing the global structural change and local residue level details, we capture the unfolding process and identify residues that degrade first, which we refer to as the proteins' 'weak links.' Subsequently, we conjugate the proteins with PEG to understand how PEG helps preserve the protein structure at higher temperatures. Results from this work will help us design novel polymer conjugates customized to specific proteins for therapeutic applications.

107e Developing Optimal Hyper-Elastic Liposomes for Drug Delivery: Insights from Molecular Simulations

Francisco Hung

Liposomes and lipid nanoparticles are commonly used in delivery of drugs that have low stability or high cytotoxicity. While the effects of properties such as liposome size, shape and surface charge on drug delivery applications have been widely studied, liposome elasticity remains significantly underexplored. Results from a recent study [1] indicate that hyper-elastic liposomes made of DOPC (1,2-dioleoyl-sn-glycero-3-phosphocholine), where their stiffness was modulated by fillers inside the liposomes, had increased uptake by human breast cancer cells in vitro, and had significantly larger tumor accumulation in vivo compared to more rigid liposomes. These results suggest that hyper-elastic liposomes might be promising drug delivery vehicles, especially for diseases characterized by dense fibrous tissues (certain types of cancer, pulmonary or liver fibrosis), where we hypothesize more fluidic liposomes could squeeze into the fibrotic regions and efficiently deliver their cargo. In this talk, I will provide an overview of our molecular simulation studies, which aimed at fundamentally understanding how the mechanical properties of the liposomes are affected by variables such as the molecular structure of the lipids (tails, headgroups), the composition of the liposome, and the presence of embedded hydrophobic drugs.

[1] P. Guo, D. Liu, K. Subramanyam, B. Wang, J. Yang, J. Huang, D. T. Auguste and M. A. Moses, "Nanoparticle elasticity directs tumor uptake", *Nature Communications* **2018**, *9*, 130.

107f Far and Few in between: Probing Structure-Selectivity Relationship in Membranes Using Path Sampling Techniques

Amir Haji-Akbari, Brian Shoemaker, Hessam Malmir, Tiago Domingues, Omar Khalifa

The ability of semipermeable membranes to selectively impede the transport of undesirable ions and molecules is key to many applications, from desalination and gas separation to biological membrane transport. For instance, membranes that are only permeable to water and reject most other ions and molecules are used in water desalination. Designing more selective membranes requires characterizing the kinetics and mechanism of the transport of the species rejected by the membrane. The timescales associated with such processes, however, can be too long to be accessible to conventional MD simulations. Moreover, the driven nature of the underlying separation processes excludes the utilization of advanced sampling techniques that require microscopic reversibility. Finally, the separation of timescales between the transport of desirable species (such as water) and undesirable species (such as salts) will result in considerable changes in reservoir concentration throughout an MD trajectory, which can skew the kinetics and mechanism of transport in ways that are difficult to quantify and correct. Recently, we utilized [1] jumpy forward flux sampling (jFFS) [2], a path sampling technique developed in my group, and non-equilibrium MD to study pressure-driven ion transport through nanoporous graphitic membranes. Our approach addresses all these challenges. It not only allows us to accurately and efficiently estimate arbitrarily long mean first passage times in pressure- and field-driven filtration processes, but also to compute fluxes and permeabilities within a pseudo-equilibrium ensemble in which both reservoirs are at different- but almost constant- chemical potentials.

In this presentation, I will discuss several technical aspects of this new approach, particularly a new model that we have developed to correct for the rather strong finite size artifacts in simulations of non-equilibrium ion transport [3]. What is remarkable about this model is that it corrects for finite size effects "on the spot", i.e., from the information obtained from a single finite simulation. Therefore, the behavior of the system in the thermodynamic limit can be inferred without a need to conduct multiple simulations of systems with different sizes (as is the common practice in the computational chemistry community). I will also discuss some of our more recent findings on how pore topology impacts the fidelity of order parameters and how pore arrangement within a membrane can impact transport resistance and selectivity in nontrivial ways.

[1] Malmir H, Epsztein R, Elimelech M, Haji-Akbari A, *Induced charge anisotropy: a hidden variable affecting ion transport through membranes*, **Matter**, 2: 735 (2020).

[2] Haji-Akbari A, *Forward-flux sampling with jumpy order parameters*, **J. Chem. Phys.**, 149: 072303 (2018).

[3] Shoemaker BA, Domingues TS, Haji-Akbari A, *Ideal conductor model: An analytical finite-size correction for nonequilibrium molecular dynamics simulations of ion transport through nanoporous membranes*, **J. Chem. Theory Comput.**, 18: 7142 (2022).

107g Accurate and Low-Cost Predictions of Complex Solvation and Reactivity Phenomena in Chemistry and Catalysis

Alexander V. Mironenko

The transition from fossil fuels to renewable energy calls for discoveries of new catalytic materials and reaction mechanisms that will enable novel transformations linking renewable energy sources (biomass, water, carbon dioxide, solar light) and high-quality fuels and value-added chemicals. Modern computational methods provide unique atomic-level insights into chemical processes, often invisible to experimental techniques, and promise to reduce the reliance on tedious "trial-and-error" experimentation in catalyst design. The challenge lies in the high complexity of realistic computational models that comes with a steep increase in computational cost. In this talk, I describe two case studies showcasing the power of current and emerging computational methods to describe complex phenomena in heterogeneous catalysis and chemistry at low computational cost and with near-quantitative accuracy.

First, I describe a multiscale protocol for making reliable predictions of adsorption enthalpies of up to 55-atom molecules in hydrophilic (defected) and hydrophobic (pristine) microporous Lewis-acidic zeolites from dilute liquid solvent mixtures in complete agreement with calorimetry measurements. The structure of the complex environment inside the pores, observed with an atomistic resolution, is shown to influence the rates of zeolite-catalyzed reactions, such as alkene epoxidation by H_2O_2 and epoxide ring opening.

Second, I discuss how applying ideas from thermodynamics to quantum chemistry leads to a simple, highly accurate, and non-empirical tight binding-type method for making predictions of chemical reactivity at a low cost with broad implications across chemistry, catalysis, and materials science.

451a Graph Neural Networks to Predict Ordering-Dependent Properties from Unrelaxed Multicomponent Crystal Structures

Jiayu Peng, Jessica Karaguesian, James Damewood, Daniel Schwalbe-Koda, Rafael Gomez-Bombarelli

Multicomponent perovskite oxides are an essential class of inorganic materials with diverse applications, where the ordering of cations can profoundly influence their physical and chemical properties. Unfortunately, the discovery and optimization of such oxides have been significantly hindered by their compositional and structural complexity. State-of-the-art brute-force experimental or computational methods can hardly resolve such complexity efficiently. Even one multicomponent oxide composition can have tens to hundreds of cation orderings, which are too expensive for brute-force high-throughput screening.

To boost the high-throughput screening of multicomponent perovskite oxides, graph neural network models can be used to effectively predict material properties from the connections of atoms in crystal structures, allowing efficient computational discovery over a vast search space. However, while these models have been demonstrated to capture the compositional dependence of material properties across a broad chemical space, it is unclear whether such models can also effectively learn the ordering dependence of similar properties for a given composition. In addition, structures after costly density functional theory (DFT) relaxations have been dominantly used for training and leveraging such graph neural network models, which cannot truly eliminate the prohibitive cost of DFT in high-throughput screening.

In this work, we developed graph neural network models to accurately predict key cation ordering–dependent properties of multicomponent perovskite oxides from DFT-unrelaxed cubic perovskite structures. Essentially, models that infer material properties by indirectly learning the thermodynamics of mixing lead to lower errors than those that predict these properties directly. Moreover, equivariant neural networks [1,2] better capture the ordering dependence of such properties than their symmetry-invariant counterparts [3] due to their higher expressivity in distinguishing the difference between the coordination environments and symmetries of various orderings. Lastly, using contrastive losses in the training of these models further increases the prediction accuracy by facilitating an optimal balance between capturing both the compositional and structural dependence of material properties. Overall, this work highlights an effective strategy for designing machine learning models with optimized robustness and generalizability for discovering multicomponent materials.

References:

[1] Geiger, et al. arXiv:2207.09453 (2022).

[2] Schütt, et al. arXiv:2102.03150 (2021).

[3] Xie, et al. *Phys. Rev. Lett.* 120, 145301 (2018).

451b Designing Molecular Coordination Environments for Selective Ion Binding Using Machine Learning

Shuwen Yue, Aditya Nandy, Heather Kulik

The design of ion selective materials with improved separations efficacy and efficiency is critical for the development of novel membrane technologies in water purification and desalination. In material classes such as polymeric membranes, MOFs, 2D nanosheets, and carbon nanotubes, pore geometry and chemistry are highly tunable, for example through chemical functionalization. In this study, we utilize a data-driven approach to investigate the design knobs of material structure and chemistry which can enhance ion selectivity. We curate a dataset of 1022 alkali metal coordinating molecular complexes from the Cambridge Structural Database (CSD) and obtain ion binding energies from density functional theory (DFT) calculations. Our analysis reveals that energetic preferences correlate strongly with ion size and are largely due to enthalpic interactions rather than structural rearrangement. We then use machine learning models to identify significant features in ion coordination, such as coordination distances, electrostatic potential, and charge, that can discriminate between alkali ion species. These physical insights offer guidance toward the design of optimal membranes for ion selectivity.

451c Developing 3D Convolutional Neural Networks (ZeoNet) for Predicting Adsorption in Nanoporous Zeolites

Yachan Liu, Gustavo Perez, Zezhou Cheng, Aaron Sun, Sam Hoover, Wei Fan, Subhransu Maji, Peng Bai

Zeolites are nanoporous materials widely applied in the chemical industry with many potential applications in chemical separations and catalysis. In this work, we present ZeoNet, a representation learning framework using 3D convolutional neural networks (ConvNets) and a new volumetric representation, to predict Henry's constants for adsorption of long-chain hydrocarbon molecules in all-silica zeolites. The impact of various factors on model performance was systematically investigated and optimized, including ConvNet architecture, the linear dimension and spatial resolution of the volumetric grids, batch size, optimizer, and learning rate. The best-performing ZeoNet achieved a correlation coefficient $r^2 = 0.977$ and a mean-squared error MSE = 3.8 in lnkH, which corresponds to an error of only 9.3 kJ/mol in adsorption free energy. Additionally, feature attribution and visualization of trained models showed that the predictions by ConvNets are driven primarily by the accessible pore volume rather than the region occupied by framework atoms. Our study provides insights into how 3D ConvNets can best be used to represent extended materials systems as well as a collection of pretrained models for use in other applications where training samples are less abundant.

451d Providing a Robust Experimental Database Model for Organic Thin-Film Electronics

Aaron L. Liu, Rahul Venkatesh, Myeongyeon Lee, Jessica A. Bonsu, Ron Volkovinsky, Azzaya Khasbaatar, Ying Diao, Carson Meredith, Elsa Reichmanis, Martha Grover

The domain of π-conjugated polymer semiconductors is undoubtedly an active frontier for next-generation materials and a keen area of interest for both the materials informatics and organic electronics communities, due to their imminent promise in large-area, printable, deformable devices and mobilizing new energy applications. However, despite recent successes in the data-enabled development of organic transistors, photovoltaics, and electrochromic devices, rational design for organic thin-film electronics, particularly those that are polymer-based, still largely materialize through one-parameter-at-a-time, hypothesis-driven studies due to a lack of available representative experimental data. The materials domain of π-conjugated polymer semiconductors remains a research area in dire need of experimental data management solutions. However, experimental database management for polymer-based electronics is not trivial, especially when a plurality of the relevant attributes related to the provenance of the sample must be included accurately to form a robust, "reusable" data record. Fully capturing this information is not easy for organic devices. Because they are designed with critical device geometries at the nano- to micrometer length scale, their performance metrics are highly sensitive to their process history. Capturing this complex set of information requires the availability of domain standards and flexible data structures to account for the evolving process space that is inherently dynamic in experimental research. In this work, using organic field-effect transistors (OFETs) as a model system, a data model toward representation of organic device experiments is proposed, designed, and implemented in an object-relational database. To guide a robust representation of device process history, this work draws upon ISA-88, an international standard for automation in batch process control, to construct generalizable relationships across process transformations within the OFET fabrication procedure. By building a data repository that can handle the various nuances of the complex OFET design space, we present a foundation toward a suitable data-enabled platform to enhance hypothesis design, scientific decision-making, and model development within the traditionally "small data" space of the organic electronics community. Additionally, by designing a data model around an accepted process standard and defining keywords centered around the experimental procedure, this work not only enables a database design toward Findable, Accessible, Interoperable, and Reusable (FAIR) data representations for organic thin-film transistors, but also presents an adaptable prototype for both the organic electronics and the broader materials community.

451e A Graph Neural Network Approach for Efficient and Accurate Macromolecular Similarity Calculation

Jiale Shi, Debra J. Audus, Bradley Olsen

Pairwise similarity between macromolecules plays a significant role in numerous cheminformatics tasks such as ranking, clustering, and classification. Previous approaches for macromolecular similarity calculations involve generating chemistry-informed graph representations of macromolecules with nodes representing monomers and edges representing bonds. Subsequently, similarity scores are computed via graph edit distance or graph kernel methods. However, these approaches have significant limitations. Graph edit distance, which measures the minimum operation costs to transform one graph to another graph, is a nondeterministic polynomial (NP) complete problem. This limitation hinders the use of graph edit distance in time-sensitive applications, such as the real-time ranking of search results. Conversely, graph kernels transform graphs to fixed-length feature vectors and then compute an inner product of the feature vectors, yielding a pairwise similarity score. Without the need to do feature extractions, graph kernel methods offer improved efficiency, but they suffer from reduced accuracy and interpretability compared to graph edit distance.

In this study, we propose a graph neural network model to address these challenges and accelerate macromolecular similarity calculations. The first step involves creating a database of macromolecule similarity by calculating and collecting the exact graph edit distances. The graph neural network model then proceeds through four stages: (1) node-level embedding, transforming each node of a graph into a vector by extracting the nodes' feature and structural properties; (2) graph-level embedding, generating one embedding vector for each graph by aggregating node embeddings; (3) graph-graph interactions using both the neural tensor network and the pairwise node comparison; and (4) fully connected network layers for graph edit distance predictions. The graph neural network is trained using the macromolecule pairwise graph edit distance database. Our proposed graph neural network model overcomes the limitations of previous graph similarity calculation methods. It substantially reduces computational costs while maintaining high accuracy and offers an efficient and precise solution for calculating the pairwise similarity of macromolecules. This novel method represents a significant advancement in cheminformatics for macromolecules, paving the way for the development of advanced search engines and quantitative design tools in macromolecules.

451f A Data-Driven Interpretation of the Stability of Organic Molecular Crystals

Rose Cersonsky, Michele Ceriotti

Due to the subtle balance of intermolecular interactions that govern structure–property relations, predicting the stability of crystal structures formed from molecular building blocks is a highly non-trivial scientific problem. A particularly active and fruitful approach involves classifying the different combinations of interacting chemical moieties, as understanding the relative energetics of different interactions enables the design of molecular crystals and fine-tuning of their stabilities. While this is usually performed based on the empirical observation of the most commonly encountered motifs in known crystal structures, we propose to apply a combination of supervised and unsupervised machine-learning techniques to automate the construction of an extensive library of molecular building blocks. We introduce a structural descriptor tailored to the prediction of the binding (lattice) energy and apply it to a curated dataset of organic crystals, exploiting its atom-centered nature to obtain a data-driven assessment of the contribution of different chemical groups to the lattice energy of the crystal. We then interpret this library using a low-dimensional representation of the structure–energy landscape and discuss selected examples of the insights into crystal engineering that can be extracted from this analysis, providing a complete database to guide the design of molecular materials.

451g A Transferable Diffusion Model for Coarse-Grained Backmapping

Mike Jones, Kirill Shmilovich, Andrew Ferguson

Coarse-grained molecular models of proteins permit access to length and time scales unattainable by all-atom models and enable the simulation of important processes that occur on long time scales such as aggregation and folding. The reduced resolution of the coarse-grained models enables realization of computational accelerations, but sacrifices the atomistic resolution that can be vital for a complete understanding of the mechanistic details. Backmapping is the process of restoring the all-atom details to coarse-grained molecular representations in order to recover atomistic-level insight. Conventional backmapping approaches generate initial all-atom structures based on geometric rules and then apply energy relaxation to eliminate aphysical high-energy overlaps and produce stable all-atom configurations. The need for energy minimization makes these procedures typically quite expensive and slow. Recently, data-driven approaches have demonstrated great promise in furnishing trainable models to efficiently perform backmapping of small molecules and proteins. In this work, we report a novel backmapping approach based on autoregressive denoising diffusion probability models to restore all-atom details to coarse-grained simulations represented only by C-alpha coordinates. The generation process is conditioned on the coarse-grained protein configuration and any previously backmapped side chains in an autoregressive fashion in order to avoid steric clashes. As an inherently transferable and local model, it is scalable to proteins of arbitrary size with linear scaling. We train the model on over 100K proteins in the SidechainNet training data set and demonstrate state-of-the-art performance on systems including DE Shaw training trajectories of fast-folding mini-proteins, ensembles of intrinsically-disordered proteins, and randomly sampled selections from the Protein Data Bank. Furthermore, we demonstrate that fine-tuning the transferable model on a given system can further improve performance in recapitulating protein-specific sidechain distributions. We make the backmapping tool available as a free, open source Python package.

451h Informing Graph Convolutional Networks with Molecular Mechanics: A Case Study of Sigma Profile Prediction

Dinis Abranches, Edward Maginn, Yamil Colón

Sigma profiles, a type of molecular descriptor obtained through quantum chemistry calculations, have shown great performance when used as a feature for machine learning models. However, because the computation of sigma profiles involves DFT optimizations, often for reasonably large molecules, they are expensive and time-consuming to obtain. Furthermore, there are no free, open-source tools readily available to directly calculate them. As such, the development of predictive methodologies able to estimate sigma profiles without the use of quantum-chemistry-based calculations would greatly enhance the applicability of these molecular descriptors in large scale machine learning models.

Graph convolutional networks (GCNs) are a class of machine learning models that have a similar architecture to convolutional neural networks but operate on graphs. Graphs are a mathematical construction of nodes and edges, and are represented by adjacency matrices, a table indicating the connectivity between nodes. There is an obvious resemblance between the mathematical concept of a graph and molecular structures, with atoms and bonds being represented by nodes and edges, respectively. Thus, because molecules are easily translated to graphs without the need to compute quantum-chemistry-based descriptors, GCNs are the ideal candidate to accelerate the development and collection of vast sigma profile datasets.

A well-known issue of GCNs pertains to the accurate description of the global and local environment of atoms in molecules, which often requires many convolution iterations, leading to over-smoothing and poor model performance. For example, the chemical natures of aromatic and aliphatic carbons are markedly distinct, but several iterations would be necessary for a GCN to learn a proper graph embedding of a ring with six nodes (e.g., benzene ring). Fortunately, these issues have been addressed at length by the molecular mechanics community. That is, commonplace to the development of classical force fields is the necessity of distinguishing the polarity and behavior of atoms with different neighboring environments. Thus, force fields typically assign class types (i.e., numerical values) to distinguish between atoms with the same atomic number but different neighboring moieties. The existence of this type of atom classification, developed throughout decades of trial-and-error methodologies and heuristics, may prove to be an excellent type of node-level featurization to be used in GNNs and, thus, merits its investigation.

Motivated by the need to expedite the development of large sigma profile datasets, the objective of this work is to ascertain the usefulness of force field atom types as node-level features in GNN models. To do so, graph representations were constructed for the compounds present in a freely available sigma profile database, with three types of node-level features being explored: simple atomic numbers, general Amber force field (GAFF) atom types, and Merck molecular force field (MMFF) atom types. GAFF and MMFF were chosen due to the breadth of molecules that both force fields can simultaneously describe, as well as the availability of tools that perform their atom typing automatically. Combining these node-level features with GCNs led to models with excellent sigma profile predictive capability (coefficients of determination larger than 0.95), with both force fields displaying similar performances. Finally, to demonstrate their practical applicability, the sigma profiles predicted using this methodology were used as the input to machine learning models previously developed in the literature to estimate the boiling temperature and aqueous solubility of organic compounds, yielding performances that rival those attained using sigma profiles obtained through quantum chemistry calculations.

451i Characterizing the Multi-Scale Structure and Dynamics of Soft Gels through the Euler Characteristic

Alexander Smith, Emanuela Del Gado, Victor Zavala

Soft gels have many mechanisms for formation, including phase separation, hierarchical aggregation, and self-assembly of soft matter materials (e.g., polymers, colloids) [1,2]. The stochastic and multi-scale nature of these mechanisms causes soft gels to have hierarchical structure. This structure imbues soft gels with physical properties, such as texture, elasticity, and stability [1,2,3]. These diverse mechanical properties can be optimized for a broad range of industrial applications such as cosmetics, building materials, and pharmaceuticals [4,5,6].

A major bottleneck in the design and application of soft gel materials is the lack of fundamental understanding of the link between the gel's microstructural topology and geometry and their physical characteristics [1,2]. This is difficult to understand due to the multiple phenomena observed in soft gels such as stiffening, aging, and yielding that occur across time and length scales within the gel [3,7]. These phenomena are described by dynamic changes in both topology (bonding network) and geometry (spatial distribution) of the material [7]. Furthermore, many of the methods used in understanding these phenomena focus exclusively on either the geometry or topology of soft gels which does not provide a wholistic characterization.

In this presentation, we use the Euler characteristic to quantify both the geometry and topology of soft gel molecular simulations [1]. The Euler characteristic (EC) is a simple, interpretable descriptor of the shape of data [8,9]. The EC captures and quantifies the topological invariants of a shape, such as the number of connected components, cycles, and voids. Furthermore, these topological invariants are directly related to the geometry of the gel, allowing us to capture both topology and geometry simultaneously. We couple our EC analysis with a filtration, a method developed in the field of Topological Data Analysis, which we use to quantify the shape of our soft gel simulations at multiple scales [8,9]. This technique also allows us to bypass the need for parameter selection, improving the robustness of our analysis in comparison to other parametric methods.

We leverage the EC to capture and differentiate the topological and geometrical changes of soft gels undergoing shear stress. The computation of the EC is easily scaled to large simulated systems, allowing us to analyze hundreds of simulation snapshots with tens of thousands of molecules in minutes using common computational hardware. Through the EC, we identify significant phase changes in gel multi-scale topology/geometry as the gel is stiffened and when the gel undergoes yielding. These results provide insight into the multi-scale mechanisms associated with nonlinear responses of gels, which are important for understanding and designing the dynamic behavior of gels to optimize hardening, stiffening, or self-healing [3]. We also explore the physical connections between the observed topological/geometric phase transitions and the characteristics of the resulting soft gel.

References:

[1] Colombo, Jader, and Emanuela Del Gado. 'Stress localization, stiffening, and yielding in a model colloidal gel.' Journal of rheology 58.5 (2014): 1089-1116.

[2] Bouzid, Mehdi, and Emanuela Del Gado. 'Network topology in soft gels: Hardening and softening materials.' Langmuir 34.3 (2018): 773-781.

[3] Bouzid, Mehdi, and Emanuela Del Gado. 'Mechanics of soft gels: Linear and nonlinear response.' Handbook of Materials Modeling: Applications: Current and Emerging Materials (2020): 1719-1746.

[4] Burey, Paulomi, et al. 'Hydrocolloid gel particles: formation, characterization, and application.' Critical reviews in food science and nutrition 48.5 (2008): 361-377.

[5] Gallegos, C., and J. M. Franco. 'Rheology of food, cosmetics and pharmaceuticals.' Current opinion in colloid & interface science 4.4 (1999): 288-293.

[6] Masoero, Enrico, et al. 'Nanostructure and nanomechanics of cement: polydisperse colloidal packing.' Physical review letters 109.15 (2012): 155503.

[7] Schall, Peter, David A. Weitz, and Frans Spaepen. 'Structural rearrangements that govern flow in colloidal glasses.' Science 318.5858 (2007): 1895-1899.

[8] Smith, Alexander, and Victor M. Zavala. 'The Euler characteristic: A general topological descriptor for complex data.' Computers & Chemical Engineering 154 (2021): 107463.

[9] Smith, Alexander, et al. 'Topological Analysis of Molecular Dynamics Simulations using the Euler Characteristic.' Journal of Chemical Theory and Computation (2022).

451j Order Parameter-Free Classification of Complex Local Environments of Shape Particles

Shih Kuang Lee, Sun Ting Tsai, Sharon C. Glotzer

Detecting and analyzing the local environment is crucial for investigating the dynamical processes of crystal nucleation and shape colloidal particle self-assembly. Recent developments in machine learning provide a promising avenue for better order parameters in complex systems that are challenging to study using traditional approaches. However, the application of machine learning to self-assembly on shape particles is still underexplored. To address this gap, we propose a simple yet powerful approach that involves training a multilayer perceptron (MLP) as a local environment classifier for shape particles, using input features such as particle distances and orientations. Our MLP model is trained in a supervised manner with shape symmetry-encoded data augmentation technique, without the need for any traditional order parameters or symmetry functions. We evaluate the performance of our model on four different scenarios involving self-assembly of cubic structures, 2-dimensional and 3-dimensional patchy shape particle systems, hexagonal bipyramids with varying aspect ratios, and truncated shapes with different degrees of truncation. The proposed training process and data augmentation technique are both straightforward and highly interpretable, enabling easy application of the model to other processes involving particle orientations. Our work thus presents a valuable tool for the investigation of self-assembly processes on shape particles, with potential applications in structure identification of molecular or coarse-grained systems.

451k Evaluating Robustness of Machine Learned Force Fields with Enhanced Sampling Methods

Yezhi Jin, Gustavo Perez Lemus, Pablo Zubieta, Juan J. de Pablo

Molecular Dynamics simulations have become a essential part in materials science and engineering. Machine Learning force fields have emerged as a useful tool to achieve ab initio accuracy in energy and forces while maintaining the speed of classical simulations. However, validating the potentials based only on force and energy accuracies is insufficient as system stability is often overlooked. This study evaluates the robustness of different machine learning models such as DeePMD, Graph Neural Network force fields, and Gaussian Approximation Potential (GAP) by performing molecular dynamics simulations using PySAGES, a python library for advanced sampling simulations, with ASE or JAX-MD backends. The study examines the stability of different systems and evaluates free energy landscapes as a function of appropriate collective variables. The results demonstrate that some models exhibit undesired behavior such as bond breaking, while others maintain the most relevant set of features that are needed to take advantage of them as widely used force fields. PySAGES is shown to be a useful tool for quickly evaluating the soundness of ML force fields.

545a Expanding Density-Correlation Machine Learning Formalisms for Anisotropic Particles and Hierarchical Systems

Rose Cersonsky

In recent years, machine learning (ML) methods have transformed computational chemistry and materials research. At the start of any machine learning inquiry we must first ask: how do we represent our system? This question is at the core of many ML methods, ranging from end-to-end methods, that take 'raw' data formats and use multi-layer architectures to implicitly encode the chemical information to 'feature-forward' methods, that use physics and chemistry to design numerical representations conducive to machine learning workflows. While both approaches have merits and overlaps, the latter often allows for added tunability to identify the role of key parameters in the design space and more straightforward interpretation.

There are many ways to encode the raw chemical data into features, and the suitable choice largely depends on the problem at hand. For cheminformatics, string-based featurizations such as SMILES or SELFIES are popular, as they encode important parameters such as present functional groups and connectivity and can be parsed using NLP models and other text-based advancements. However, in thermodynamic contexts, where the chemistry and connectivity remains largely unchanged, such as in molecular simulation, it is more typical to use configuration-dependent features, which transform molecular coordinates into a range of suitable numerical representations.

Yet, that leaves many of us that work beyond the atomic scale without many options for thermodynamic representations that reflect the complex interactions of mesoscale building blocks. Of the frameworks developed for atomic representations, density-based frameworks present the most compelling avenue for expansion. Furthermore, by defining our representations within the same conceptual framework as our atomic counterparts, we beget an ability to combine or compare representations across multiple scales, which is desperately needed for multiscale simulation and understanding and designing cross-scale phenomena.

Here, we introduce and demonstrate a first-such expansion of density-based frameworks for ML representations by taking the popular SOAP (Smooth Overlap of Atomic Positions) formalism and demonstrating its expansion to simple anisotropic bodies. We provide both the mathematical theory behind this expansion and demonstrate several case-studies for its usage across multiple simulation length-scales and materials systems.

545b Accelerating Molecular Dynamics with Field-Theoretic Simulations

Joshua Lequieu

Particle-based simulations like molecular dynamics are ubiquitous throughout academia and industry, yet these methods suffer from several well-known limitations. One limitation is that particle-based simulations struggle to reach equilibrium when a system contains a broad spectrum of timescales, such as in polymeric materials, which often necessitates excessively long simulation times. Another limitation is that the free energy is not directly accessible in a particle-based simulation and must be computed using enhanced sampling methods. In this talk, I describe a new method that address both of these limitations. The basis for our approach is a new type of ``multi-representation" simulation where particle and field-theoretic simulations are linked together into a unified framework. Our approach involves the construction of formally equivalent particle and field-theoretic models that are simulated subject to the constraint that their spatial density profiles are equal. This constraint provides the ability to rapidly map between particle and field-theoretic representations so that simulations can be performed in the particle/field representation where calculations are most efficient. Our approach can leverage the superior numerical properties of field-theoretic simulations to rapidly relax the largest length scales within a system and to calculate the free energy, yet can backmap into a particle-based representation when molecular configurations and dynamics are desired. Furthermore, by leveraging the formal equivalence between particle and field-based simulations, these calculations can be placed on a rigorous theoretical foundation that can be constructed without approximation. Our method is illustrated using both linear diblock copolymers and intrinsically-disordered proteins. In both cases, our method can provide direct access to the free energy, rapid equilibration, molecular configurations and dynamics.

545c Generative Model Assisted Sampling of Multiscale Dynamical Systems

Ellis R. Crabtree, Juan Bello-Rivas, Andrew Ferguson, Ioannis G. Kevrekidis

Sampling the phase space of molecular systems -- and, more generally, of complex dynamical systems effectively modeled by stochastic differential equations -- is a crucial modeling step in many applications, from protein folding to materials discovery. These problems are multiscale in nature and can often be described in terms of low-dimensional effective free energy surfaces parametrized by a small number of “slow' reaction coordinates; the remaining “fast' degrees of freedom populate an equilibrium measure on the reaction coordinate values. Over the years, enhanced sampling techniques coupled with molecular simulation have been developed. An intriguing analogy arises with the field of Machine Learning (ML), where generative models such as Score-Based Diffusion Models (SGMs) and Generative Adversarial Networks (GANs) can produce high dimensional samples from probability distributions parameterized by a relatively small number of degrees of freedom. This sample generation returns plausible high dimensional space realizations of a model state, from information about its low-dimensional representation. We present an approach that couples physics-based simulations and biasing methods for sampling conditional distributions with ML-based conditional generative models for the same task. The “coarse descriptors' on which we condition the fine scale realizations can either be known a priori or learned through nonlinear dimensionality reduction.

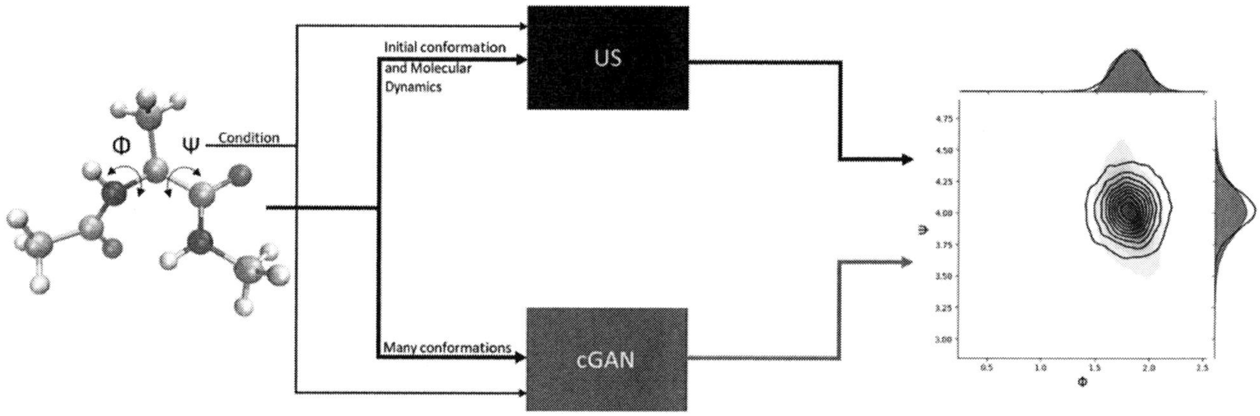

545d Multi-Scale Approach to Predict Tunable Organic-Inorganic Nanoparticle Self-Assembly

Xin Qi, François Baneyx, Brandi Cossairt, Jaehun Chun, Christopher J. Mundy, Jim Pfaendtner

Inspired by naturally occurring hierarchical materials, numerous beautiful and functional mesoscale architectures have been constructed in labs using solid-state nanoscale building blocks and bio- or biomimetic molecules via self-assembly. Their structures and assembly dynamics are tightly regulated by the system's convoluted energetics and entropic information, which may even respond to a minuscule change. For example, we recently reported a pH-regulated self-assembly of silica nanoparticles facilitated by an engineered bifunctional silica-binding protein sfGFP::Car9-Car9. The assembly and disassembly can be reversibly and cyclically modulated by alternating the solution pH between 7.5 and 8.5. In another study, we observed that an increment of the peptoid (i.e., the ligand) length by two monomers at a time would change the arrangement of a CdS quantum dot superlattice from close packing to square, further to monoclinic. However, the complexity of the nanoscale colloidal system often makes it difficult to probe the underlying physics via experimental means. Alternatively, we can construct suitable multi-scale theoretical toolkits by carefully selecting and combining theories and simulation algorithms from different levels (i.e., from ab initio to coarse-grained models) to connect the scales and gain physical insights into the assembly mechanisms. In this talk, I will elaborate on how we used multi-scale theoretical frameworks to reveal the key driving forces of self-assembly in these two cases. The knowledge we gained on the distinctive roles of organic ligands has further inspired new design principles in future studies

545e Finite Element Modeling of Atmospheric Water Extraction By Way of Highly Porous Adsorbents: A Roadmap for Solver Construction with Model Factor Sensitivity Screening

Evan Gildernew

A Finite Element Model (FEM) is developed for use in determining adsorption system performance. The model is intended to guide novel adsorbent structure fabrication and atmospheric water harvesting device design. We survey a variety of governing equation factor inputs and relationships which describe the interaction between Zeolite 13X and water vapor. Mitigation strategies are discussed for detecting the breakdown of continuum modeling at micro scales wherein Knudsen effects and other anomalous behaviors emerge. Characterization of model factor inputs and the techniques for their sourcing is described with consideration to the construction of a high throughput multi-scale shape optimized computational schema. Four objectives guided the development of this model. Our first objective was to understand the implementation of adsorption systems equations and the assumptions that could prevent reliable predictability. The second objective was to assemble, reduce, and analyze model constants and approximations that express FEM coefficient calculations as physical forces and thermodynamic properties which could be derived from other computational methods. Third, to analyze factor sensitivity of model inputs by way of a 2^k factorial screening to determine which inputs are driving the physics of water harvesting adsorption systems. The fourth objective was to design the FEM solver for integration into a multi-scale high throughput topologically optimized schema. The main finding of the solver factor screening indicate that total micropore volume is highest value characteristics in relation to water uptake.

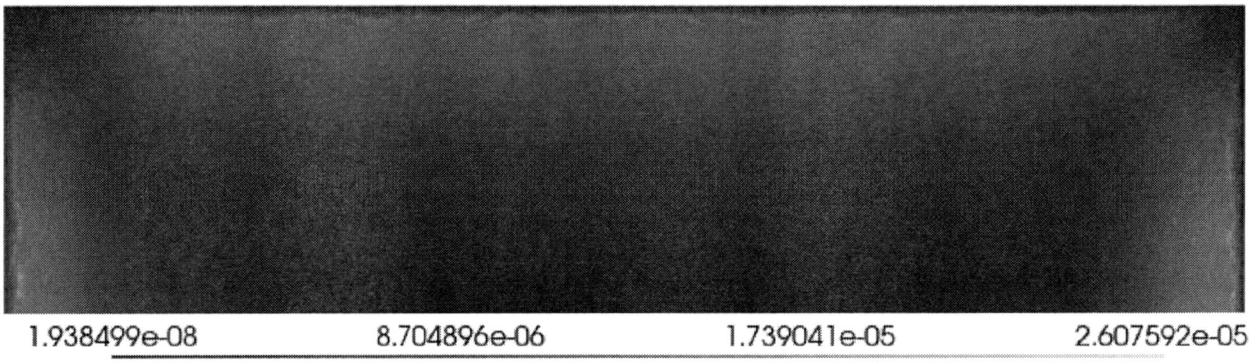

545f Multiscale Simulations of Monoclonal Antibodies in High Concentration Formulations

Harold Hatch, Christina Bergonzo, Violetta Burns Casamayor, Guangcui Yuan, Alexander V. Grishaev, Yun Liu, Vincent K. Shen

Although monoclonal antibodies (mAbs) are some of the most profitable and promising pharmaceuticals for targeted therapies, physical instabilities at high concentration including aggregation, high viscosity and phase separation cause problems for their manufacture, delivery to patients and long term stability. In this work, we present a multiscale methodology which uses all-atom modeling and experimental second osmotic virial coefficients to develop coarse-grained models for flat-histogram Monte Carlo simulations of hundreds of mAbs with sub-nanometer resolution. In this multiscale modeling approach, the major assumption is that mAb domains are held fixed so that their atomistic interactions with implicit solvent can be precomputed and therefore increase simulation efficiency by a few orders of magnitude. Although domains within the mAbs are rigid, the coarse-grained model includes the flexibility of the hinge region, which is often neglected but here shown to play an important role at mAb concentrations up to 150 g/L. This approach is also amenable to modeling excipients, co-formulations and surface interactions. Simulations were validated against experimental measurements and used to predict the physical stability of mAbs. These results highlight the potential for this multiscale approach to pre-screen pharmaceutical candidate mAbs in early stage development to avoid high concentration physical instabilities that can plague later stage development.

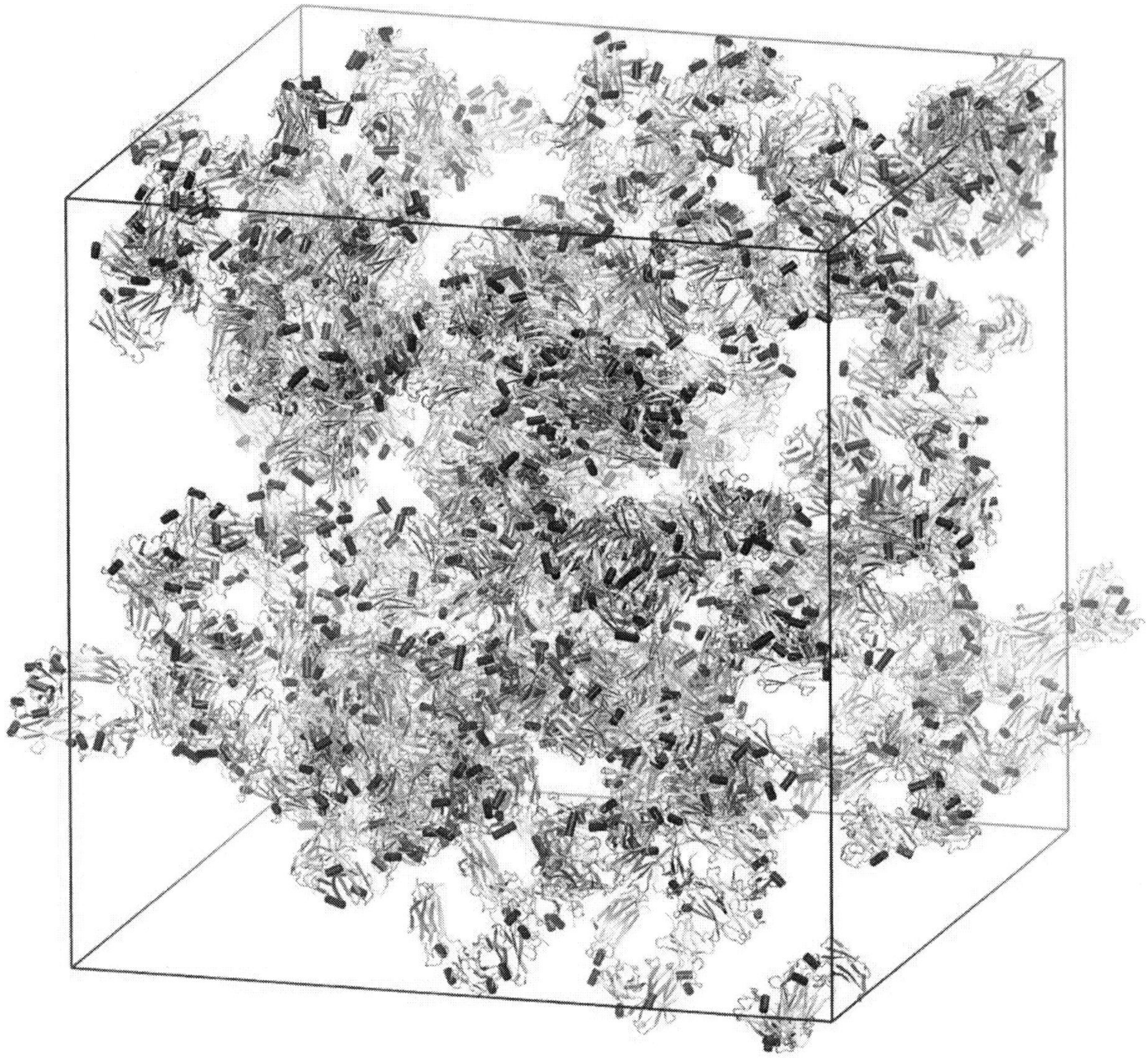

545g Flexible Backbone Membrane-Associated Protein Docking Improves Its Performance on Expanded Transmembrane-Protein Complex Datasets.

Rituparna Samanta, Ameya Harmalkar, Jeffrey J. Gray

The oligomerization of protein macromolecules on cell surfaces plays a fundamental role in regulating cellular function, including signal transduction and the immune response. Despite their importance, membrane proteins (MPs) represent only 2% of all protein structures in the protein data bank (PDB), and their complexes are even scarcer. Computational methods are promising tools for modeling MP interfaces and predicting complex structures. Here, we present RosettaMPDock, a flexible transmembrane protein docking protocol that captures binding induced conformational changes. To generate diversity in backbone conformations for the RosettaMPDock, we used three conformer generation methods: perturbation of the backbones along the normal modes by 1 Å, refinement using the Relax protocol, backbone flexing using the Rosetta Backrub protocol. RosettaMPDock samples large conformational ensembles of flexible monomers and docks together protein targets within an implicit membrane environment. To improve the scoring efficiency, we have used a combination of low-resolution Motif Dock Score and membrane based high-resolution score Franklin2023. RosettaMPDock is benchmarked on 30 transmembrane-protein complexes of variable flexibility dataset. Our results show RosettaMPDock successfully predicts the correct interface (success defined as achieving 3 near-native structures in the 5 top-ranked ones) for 67% of moderately flexible targets (unbound-bound backbone motion within 1.5-2.5Å) and 60% of the highly flexible targets (unbound-bound backbone motion greater than 2.5Å), a substantial improvement from the existing membrane protein docking methods. We have also developed a hybrid protocol that refines AlphaFold-multimer structures with RosettaMPDock and further improves prediction success rates from 64% to 73%.

545h Multiscale Modeling of Stratum Corneum Lipids to Understand the Barrier Function of Skin

Chloe Frame, Parashara Shamaprasad, Christopher Iacovella, Annette Bunge, Clare McCabe

The barrier function of the human skin is primarily located in the stratum corneum (SC), which is composed of corneocytes surrounded by a dense, lamellar lipid matrix made up of ceramides (CERs), cholesterol (CHOL), and free fatty acids (FFAs) of various lengths. While experimentalists can reproduce the lipid organization of intact SC with model systems using synthetic lipids,[1] the role of each lipid in the formation of the molecular structure of the SC, and hence the barrier properties of skin, is not well understood due to the complex nature of the SC lipid matrix. To address this need, atomistic molecular dynamics (MD) simulations of pre-assembled lipids in primarily bilayer configurations have been used to study the behavior of the SC lipids.[2] However, such simulations can suffer from configurational bias due to low lipid mobility, and further, simple bilayers cannot sufficiently represent the complexity of SC multilayer structures. Thus, in order to emulate the experimental model membrane formation at long timescales and large system sizes, we developed computationally efficient coarse-grained (CG) models using the Multi-State Iterative Boltzmann Inversion (MS-IBI) method.[3] Here, we begin our multiscale strategy of understanding SC lamella with the CG self-assembly of large complex multilayer mixtures of multiple CERs, CHOL, and FFAs that match previously examined experimental compositions. Specifically, we go beyond the commonly studied CER NS to simulate mixtures with CERs NS, NP, AS, and AP as well as understanding the role of FFA tail length distribution. Atomistic configurations generated by back-mapping the self-assembled multi-lamellar CG configurations provide insights into lipid organization and structure of experimental model membranes. This multiscale approach accumulates simulation and collaborative experiments providing a holistic picture of structural properties, hydrogen bonding capabilities, and neutron scattering studies to better understand the relationship between composition of SC lipids and the barrier function of skin.

References

(1) Bouwstra, J. A.; Ponec, M. The Skin Barrier in Healthy and Diseased State. *Biochimica et Biophysica Acta - Biomembranes*. **2006**. https://doi.org/10.1016/j.bbamem.2006.06.021.

(2) Shamaprasad, P.; Frame, C. O.; Moore, T. C.; Yang, A.; Iacovella, C. R.; Bouwstra, J. A.; Bunge, A. L.; McCabe, C. Using Molecular Simulation to Understand the Skin Barrier. *Prog. Lipid Res.* **2022**, *88*. https://doi.org/10.1016/j.plipres.2022.101184.

(3) Moore, T. C.; Iacovella, C. R.; McCabe, C. Derivation of Coarse-Grained Potentials via Multistate Iterative Boltzmann Inversion. *J. Chem. Phys.* **2014**. https://doi.org/10.1063/1.4880555.

545i Separation of Enthalpic and Entropic Contributions to Coarse-Grained Potentials for Resolving Dynamics Using the Iterative Boltzmann Inversion (IBI) Method

Lilian Johnson, Frederick Phelan Jr.

In recent work [J. Chem. Phys. 154,084114 (2021)], we studied a CG model that aims to preserve both chemical specificity and dynamics of a polymer melt. We used iterative Boltzmann inversion (IBI) to parameterize the conservative potential and reproduce the structural distributions of the melt from all-atom (AA) reference simulations, and Langevin dynamics to introduce a dissipative potential as a means to recover the AA dynamics. Using this method, we found we could repair the dynamics of various diffusive measures with about a 20% variation or less in the friction parameter. Here we further our analysis this CG model by examining the influence of Langevin friction on the dynamics of the individual potentials. Various dynamic correlation functions are used to quantify and compare the dynamics of the bonded and pair potentials for the diffusion corrected CG models and the AA reference systems. This leads to an integrated approach for recovering both structure and dynamics within the scope of an IBI based parametrization by properly accounting for separate enthalpic and entropic contributions to the potentials. The effect of the refined potentials on property predictions are analyzed.

545j Developing a Bottom-up Coarse-Grained Model for Sequence-Specific Polypeptoids

Daniela Rivera Mirabal, Sally Jiao, Evan Pretti, Audra DeStefano, Shawn Mengel, Rachel Segalman, M. Scott Shell

Polymer synthesis has become increasingly sophisticated and current developments in precise sequence-controlled polymers allow for new design opportunities. However, the enormous number of possible sequences requires robust and efficient modeling to understand and predict how sequence impacts material properties. Here, we use sequence-specific polypeptoids (a biomimetic of polypeptides) as a platform for developing design rules for relating chemical sequence to polymer conformation. Polypeptoids are particularly useful in this context due to their lack of backbone hydrogen bonding, isolating the effect of sidechain chemical sequence on polymer chain shape. Moreover, they are routinely synthesized at gram scale, sequence-specifically, with hundreds of different side chain functionalities, allowing for detailed experimental investigation and validation. Our earlier simulation studies of small polypeptoid systems with advanced sampling molecular dynamics methods examined changes in the local and global structure of short chains in response to the number and location of the hydrophobic and chiral monomers, with excellent agreement with experiments. However, to study broader chain shape effects and self-assembly behaviors, access to longer and multiple peptoid systems requires coarse-grained simulations that can exceed the limitations of atomistic simulations.

Here, we develop a bottom-up coarse-grained peptoid model using the relative entropy approach to create a library of peptoid monomers suitable for studying CG models of a wide range of sequences in both long chain and self-assembly simulations. We validate our CG models with experimental end-to-end distance measurements measured from double electron-electron resonance (DEER) spectroscopy. We also leverage inverse design methods to suggest sequences with unique folding and self-assembly properties into various structures as a way to investigate fundamental limits of sequence design. These new computational methods provide molecular insight into the driving forces for polymer conformation and are exciting as new in silico screening tools to guide the development of sequence-specific polymeric materials with tunable properties.

197a Fusogenic Liposome Formation: A Coarse-Grained Molecular Dynamics Simulations

Tibo Duran, Peter Ung, Jane He, Marta Wells, Shinji Takeoka, Jing Li

Purpose:

Liposomes are vesicles composed of a lipid bilayer and/or a concentric series of multiple bilayers commonly used as carriers for drug delivery purposes. In contrast to the conventional pathway of cell uptake, previous studies have proposed that fusogenic liposomes have the advantages of fusing with the cell membrane and directly releasing the cargo to the cytosol. However, there is a lack of deep understanding of the structural impact of the individual lipid on its assemblies, and uncovering the driving force that differentiates the route of the cellular uptake has been challenging. Here, we implemented coarse-grained molecular dynamics (CG-MD) simulations to predict the physicochemical properties of new lipids, simulate their propensity to self-assemble into liposomes, and determine the correlation between liposome size and lipid structural properties.

Methods:

The CG-MD simulations were implemented using the GROMACS software package with the MARTINI 3.0 force fields, and the trajectory analyses were conducted using MDAnalysis and PackMem. The all-atom model of examined new lipids, which composed of an amino acid-based head group, a spacer group, and two alkyl tail groups of various lengths, was first constructed using the webserver LigParGen. The new lipid model was then mapped and parameterized into the coarse-grained model. The constructed new CG lipid model was energy-minimized with the steepest descent algorithm, followed by a 10 ns equilibration in an isothermal–isochoric ensemble at ambient temperature. The cubic simulation system of various sizes (10 nm^3, 25 nm^3, 50 nm^3, and 100 nm^3) was built as mixture of water and randomly distributed lipids. The 200-ns simulation production runs were performed in an isothermal-isobaric ensemble using the Nosé–Hoover thermostat and the Parrinello–Rahman barostat at a pressure of 1 bar. The simulations were carried out using periodic boundary conditions in a 20-fs time step and the final trajectory was extracted for further physicochemical property analysis. All computations were performed using the High-Performance Computing (HPC) Supercomputer Cluster at Genentech.

Results:

The CG-MD simulations indicate that the examined lipids self-assemble into liposomes and micelles of various shapes and sizes within 100ns into the production run. Analysis of the radius of gyration showed that all liposomes formed were initially unstable and experienced significant fluctuations in their morphology. During this early stage, the liposomes usually transitioned from an elongated shape to an elliptical shape, eventually stabilizing into a mostly spherical shape that persisted for the duration of the simulation. Furthermore, the simulations showed that the thickness of fusogenic liposome bilayer was influenced by only certain types of amino acid head group, while the length of the spacer and the alkyl tail groups influence the liposome bilayer thickness independently from the head group. The curvature analysis provided detailed insights into the behavior of the membrane bilayer, revealing that the inner monolayer was more curved than the outer monolayer. The evaluation of three-dimensional packing defects in the membrane bilayer showed that liposome formation in this study was characterized by more deep defects than shallow defects, which indicates the regions where the hydrophobic lipid tails are transiently exposed to the aqueous polar environment leading to unstable liposomal structure. The simulation findings will be further validated through experimental studies using the same fusogenic lipids.

Conclusions:

While the simulation exercises did not accurately replicate the experimental particle size, potentially due to computational limitations, they still effectively uncovered the mechanism behind the formation of liposomes and their structural properties using newly designed fusogenic lipids with varying head groups, spacer lengths, and tail lengths. The MD computational platform developed in this study has proven to be a powerful tool that can help to facilitate the synthesis and testing of fusogenic lipids with specific desired properties. Not only this, but it is also capable of guiding the interpretation, explanation, and direction of experimental studies in this area.

Keywords:

Fusogenic liposomes, Molecular dynamics simulations, Coarse-grained, Membrane defects, Membrane curvature.

Figure 1. CG-MD fusogenic liposome self-assembly (left) and cross-sectional view (right).

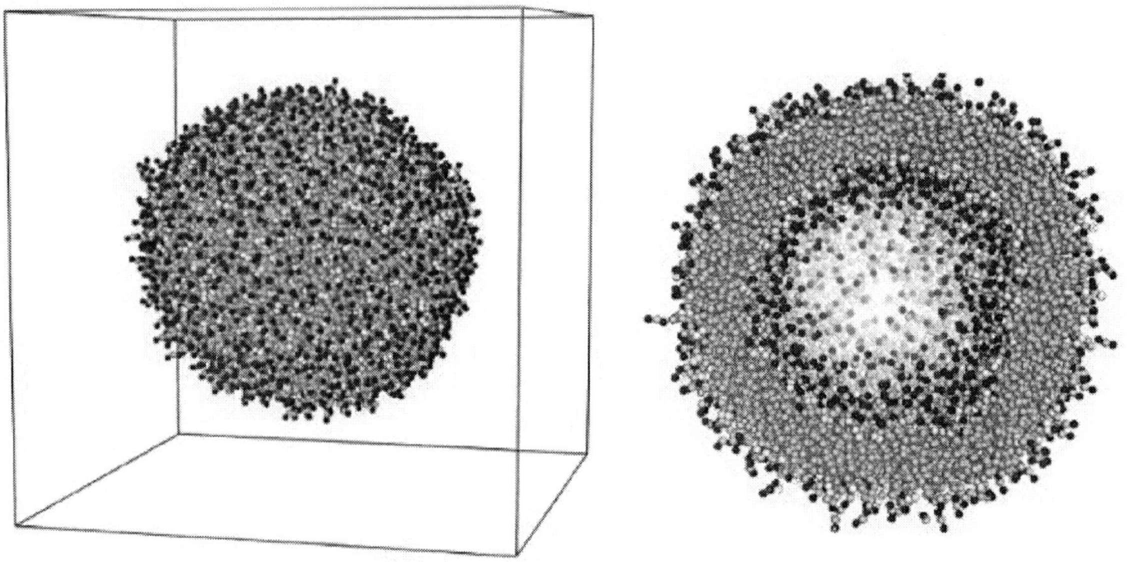

197b Using Density Functional Theory Calculations to Identify Efficient Catalysts for Dehydrogenation of LOHC

Jingwen Zhou, Sung Gu Kang

In this study, we theoretically identified the potential dehydrogenation catalysts of Pt/M/Pt(111) (M = Ni, Cu, or Pd) for one of the widely used LOHCs, perhydro-dibenzyltoluene (H18-DBT). Various strategies were applied to enhance the catalytic performance of the designed system. In particular, the structural and chemical effects on the hydrogen release from H18-DBT were systematically explored. Most interestingly, the dehydrogenation mechanism of H18-DBT over each catalyst was explored. Also, the effective descriptor to efficiently find the promising catalyst was found in this study.

197c ms2: A Molecular Dynamics and Monte Carlo Simulation Engine

Simon Stephan, Hans Hasse, Jadran Vrabec

The molecular simulation engine ms2 [1,2,3,4] is presented. It is designed for the calculation of thermophysical properties of bulk phases – in particular fluids. ms2 features the two main classical molecular simulation techniques: molecular dynamics (MD) and Monte-Carlo. It supports the calculation of vapor-liquid equilibria of pure fluids and multi-component mixtures described by rigid molecular models. For calculating phase equilibria, different methods are implemented in ms2, such as the grand equilibrium method. Furthermore, it is capable of sampling various classical ensembles and yields numerous thermodynamic properties such as the Helmholtz energy and its derivatives. To evaluate the chemical potential, Widom's test particle method and thermodynamic integration are implemented. Further entropic properties can be obtained using Krikwood-Buff integration method. Transport properties can be determined by equilibrium MD simulations following the Green-Kubo formalism. ms2 is written in Fortran90 and optimized for a fast execution on a broad range of computer architectures, spanning from single processor PCs over PC-clusters and vector computers to high-end parallel machines. The standard Message Passing Interface (MPI) is used for parallelization and ms2 is therefore easily portable onto a broad range of computing platforms. The accuracy and reliability of ms2 has been shown for a large variety of fluids in preceding work.

References:

[1] S. Deublein et al., Comp. Phys. Comm. 182 (2011) 2350-2367.
[2] C. Glass et al., Comp. Phys. Comm. 185 (2014) 3302-3306.
[3] G. Rutkai et al., Comp. Phys. Comm. 221 (2017) 343-351.
[4] R. Fingerhut et al., Comp. Phys. Comm. 262 (2021) 107860.

197d Tuning Lobed Colloid Design to Achieve Porous Morphologies

Brunno C. Rocha, Harish Vashisth

The self-assembly of colloids is emerging as a promising approach for designing new materials with applications in engineering and biomedical fields. A class of colloidal particles termed lobed colloids can be designed with protrusions (lobes) on their surfaces to provide directionality to interparticle interactions and enhanced excluded volume to facilitate porous self-assemblies. I will present results on self-assembly of several systems of lobed colloidal particles in which different characteristics of these particles, including their shapes, the sizes of their lobes, and the nature of their interparticle interactions, are varied. The systems were studied in either monodisperse (where all particles have the same shape) or polydisperse conditions (binary mixtures of particles with different shapes). The results reveal that the morphology and porosity of the self-assembled structures can be tuned by optimally designing the shapes and interparticle interactions among lobes. The relative size of the central seed particle and the lobes attached to it characterize the seed to lobe size ratio, which is found to be a key shape design parameter to achieve specific morphologies. Moreover, the polydisperse systems formed structures characterized by higher porosities in comparison to the monodisperse systems, which indicates that the composition of the system is also an important design feature for optimal self-assembly into desired morphologies.

197e Deep Learning Architecture for Peptide Property Prediction

Daniel Garzon, Emily Baum, Kendall Sano, Camille Bilodeau

Peptides have a wide range of applications and predicting their properties is important for optimizing their use in various fields such as separation materials for chromatography. Existing approaches to peptide property prediction have limitations and do not scale well to larger molecules or take advantage of important structural features of peptides.In this project, we develop a new deep learning architecture that captures important structural features of peptides, such as hydrogen bonding motifs, to predict their properties. We curate a high-quality dataset of peptide structures and their properties for training and validation of the model, we also evaluate the performance of the model and compare it to existing methods to identify key structural features that contribute to peptide properties. Our method has the potential to significantly advance the design of peptides with optimized properties for a range of applications, future works includes expanding the scope of the model to predict other complex peptide properties and improving the accuracy of the predictions.

197f Decoding Polymer Behavior: The Power of Molecular Modeling

Feranmi Olowookere, C. Heath Turner, Jason Bara, Ali Alshaikh

Polymers are widely used in various industries for their diverse properties, but their limited recyclability raises concerns about their sustainability. Conventional recycling methods are challenged by product quality and cost, highlighting the need for alternative upcycling processes to convert these polymers to value-added products. To improve such processes, efficient solvents must be designed, which is where molecular modeling comes in. Molecular dynamics (MD) is a complementary tool to experimental studies that provides molecular insights into polymer-solvent systems. In this poster, we showcase the power of MD by presenting various fundamental properties to describe polymer behavior in two polar organic solvents: tetrahydrofuran (THF) and dimethylformamide (DMF). These properties include end-to-end distance, radius of gyration, radial distribution functions, shape descriptors, end-to-end vector correlation function, dihedral autocorrelation function, surface area, and electrostatic potential. We investigate the effects of varying chain lengths on the structure and dynamics of polyvinyl chloride (PVC) and compare our results to experimental data. Our findings suggest that the PVC120 model accurately represents experimental data when performing atomistic molecular simulations. Overall, this research demonstrates the potential of molecular modeling to decode polymer behavior and improve upcycling processes, contributing to the development of more sustainable and efficient materials in various industries.

197g Representation Learning Approach for Mapping Protein Surface Patches

Emily Baum, Daniel Garzon, Camille Bilodeau

When designing proteins or molecules that target proteins, a common issue is that proteins may dimerize and aggregate, blocking active sites and limiting intended interactions. Hydrophobicity plays an important role in protein aggregation and dimerization, but current methods for high resolution mapping of hydrophobicity on protein surfaces are still lacking. To predict hydrophobicity, we need to understand the surface configuration of proteins. In this presentation, we develop a patch based characterization of protein surfaces through representation learning. First, we break down proteins from the PDB database into surface patches. Then we construct and train a variational autoencoder to represent patches in a vector form. Understanding similarities and differences in surface patches begins to suggest relationships between molecular features and protein function. This is a crucial step toward protein hydrophobicity mapping because it reduces the latent space of all possible proteins down to common surface patches. Mapping the constructed surface patch space allows us to more easily predict protein function as well as similarities between proteins or different regions of the same protein.

197h A Study on Supercomputing Utilization and Molecular Structure Exploration Algorithm for Accelerating Material Inverse Design

Min-Ho Suh

Inverse design is a concept in which a desired set of properties is defined first, and then a material or molecule with those specific properties is designed through a systematic process. For the past 30 years, researchers have been actively studying Computer-Aided Molecular Design (CAMD) technology in the fields of bio-chemical materials and new drug development. Recently, with the application of big data and artificial intelligence technologies, the effectiveness of this technology has been realized. The primary goal of molecular design is to create new materials with improved physical properties compared to existing ones. To achieve this, an optimization process is used to propose a set of molecular structures for new materials or groups of similar materials that are expected to exhibit excellent physical properties. This process significantly reduces the number of materials that need to be synthesized or the range of physical properties that need to be measured through experiments. To maximize the efficiency of the optimization process based on supercomputing, algorithms that allow parallel searching are applied. Molecular dynamics and density functional theory are widely used to predict various properties of a substance based on its molecular structure through calculations, provided that the molecular structure is known through computation. To perform such calculations, a combination of databases and theoretical formulas are utilized. This study covers the basic composition for developing such software, as well as an example of a simple molecular design to test and develop the prototype program through a case study on the design of CO2 absorbent materials.

197i Atomic Layer Deposition Simulation Using Stochastic Parallel Particle Kinetic Simulator

Woojin Kang, Chaeeun Lee, Won Bo Lee, Jonggeol Na

The semiconductor industry has experienced significant growth and change in recent years due to the increasing demand for smaller and more powerful devices. To meet this demand, an atomic layer deposition (ALD) process has been introduced to create thinner and more precise layers. ALD processes offer high controllability, precise thickness control, and uniform thin film formation over a large area. To better understand and enhance these processes, computational modeling and simulation methodologies are being employed. This study focuses on the simulation of the diffusion and deposition process on the wafer surface using a Stochastic Parallel PARticle Kinetic Simulator (SPPARKS) and the kinetic Monte Carlo technique. Specifically, this study examines the chemical reactions of the chemical species present on the surface and sequentially pulsing and purging the surface with Dichlorosilane and ammonia gas. Adsorption, desorption, and deterioration occur probabilistically, which confirms the expected growth of the atomic layer. The growth of the atomic layer is compared with existing literature. The ultimate objective of this study is to develop a simulation model for a specific chemical species and predict the quality and thickness of the entire wafer under varying conditions, such as different pressures and temperatures. By providing insights into the ALD process and improving the ability to predict the properties of thin films, this study could play a crucial role in the development of smaller and more powerful devices in the semiconductor industry. In conclusion, this study demonstrates the power of computational modeling and simulation in advancing the ALD process. The results provide a foundation for developing a comprehensive simulation model to predict the quality and thickness of thin films on wafers under different conditions. These insights could ultimately enable the development of more precise, reliable, and efficient thin film deposition techniques, leading to the production of smaller and more powerful devices.

197j A High-Throughput Computational Framework for Investigating Protein-Polymer Bioconjugates with Molecular Dynamics Simulations

Aidan Wegner, Ziyue Dong, Michael F. Toney, Kayla G. Sprenger

Protein-polymer bioconjugates are a class of hybrid biological molecules that strategically utilize the natural specificity of proteins and the synthetic modularity of polymers. Recently, the design and assembly of protein-polymer bioconjugates have gained much attention due to the potential applications of these bioconjugates in drug therapeutics, biomineralization, and biosensing. The conjugation of polymers to protein surfaces can enhance key properties of the biomolecule such as its structural stability and affinity for specific targets, while maintaining its natural biological functions. To optimally design and synthesize these hybrid molecules, the underlying mechanisms that influence their enhanced thermodynamic and kinetic properties must be understood at the molecular level. Computational approaches such as molecular dynamics (MD) simulations provide an excellent entry point for such studies, via the development of accurate and compatible molecular models for studying broad variations of the protein-polymer bioconjugate system.

Herein, we showcase the development of a computational framework for systematically building and testing models of protein-polymer bioconjugates that can be readily scaled for high-throughput processing. This methodological workflow encompasses various *in silico* tools such as Python, C++, CGENFF, Gaussian, Avogadro, ChemDraw, VMD, and GROMACS. Specifically, we built and tested this framework on bovine alpha-chymotrypsin and hen egg-white lysozyme conjugated with sulfobetaine methacrylamide through atom transfer radical polymerization (ATRP) initiators, as well as on human interleukin 12 (IL-12) conjugated with activated polyethylene glycol (PEG). The framework itself, when combined with fundamental analysis methods such as MD, will support the rational design of novel bioconjugates with improved efficacy in terms of function and stability. To this end, we conducted MD simulations of the above bioconjugates in different chemical and biological environments, namely: (1) ATRP-initiated bioconjugates interacting with silica at various temperatures, and (2) PEG-ylated bioconjugates interacting with phospholipid membranes. In the former case, we sought to elucidate structural changes in the bioconjugates that could affect their adsorption to nanoparticle surfaces in drug delivery schemes. In the latter case, we sought to identify adsorption free energy differences between the bioconjugates and their non-conjugated counterparts to inform the use of these bioconjugates in protein therapeutics applications. Finally, through the development of this computational framework, we gained new atomistic insights into advanced protein functions and an improved understanding of the molecular driving forces governing bioconjugate stability.

197k Implementation of Genetic Algorithms to Optimize Metal-Organic Frameworks for CO_2 Capture

Thang Pham, Randall Snurr

Metal-organic frameworks (MOFs) are promising adsorbents for CO_2 capture that could be used in adsorption processes that are potentially less energy intensive than current methods for CO_2 capture using liquid amines. MOFs are highly versatile, and there are an almost unlimited number of MOFs that could be synthesized. A major challenge is to rapidly identify the most promising MOF materials for various application. In this work, we used genetic algorithms (GA) and grand canonical Monte Carlo (GCMC) simulations to efficiently search for high-performing MOFs for CO_2 capture. We determined and analyzed the effects of important GA parameters, including the mutation probability, the number of MOFs per generation and the number of GA generations on the GA performance. We performed GA on-the-fly to optimize MOFs using multiple objective functions across different topologies. GA was able to determine top-performing MOFs that expanded beyond the previously known pareto fronts and reduced the cost of molecular simulations by a factor of 25.

197m Elucidating the Molecular Mechanisms By Which Amyloid-Beta Suppresses HSV-1 Infection in the Brain

Bailey Zinger, Kayla G. Sprenger

Amyloid-beta (Aβ) is derived from the neuronal transmembrane protein Amyloid Precursor Protein, following its cleavage by a series of secretases. Self-aggregation of the resulting Aβ peptides into oligomers and plaques has been strongly implicated in mediating neurodegeneration and Alzheimer's disease[1]. However, recent studies have elucidated many *beneficial* functions of Aβ, including its ability to repair holes in the blood-brain barrier, regulate synaptic properties of neurons, adopt antimicrobial peptide-like behaviors, and even exhibit antiviral behaviors towards foreign pathogens in the brain. In the latter case, for example, Aβ has been shown to sequester and remove herpes simplex virus (HSV-1)—which affects 67% of the human population—from the brain, via direct interactions with the virus' surface glycoproteins[2]. Yet, the full picture of how Aβ mediates these antiviral behaviors with respect to HSV-1, and other viral pathogens in the brain, remains unclear.

The structure of Aβ has been observed to vary between a mostly unstructured coil in solution, to a helical form after coming into contact with a cellular membrane, to a neurotoxic beta sheet-like structure after the process of self-aggregation has begun. Moreover, Aβ commonly exists as both a 40- and 42-residue peptide, the latter of which is considered to be the more neurotoxic form[3], but also the form with the more potent antiviral capabilities. As such, we hypothesized that the 42-residue Aβ peptide that resides in an unstructured or helical form would exhibit the strongest antiviral behavior, binding to HSV-1 glycoproteins in a way that directly interferes with their ability to bind to host cell receptors. To test this idea, we performed molecular dynamics (MD) simulations, in triplicate using the GROMACS engine, of Aβ in the six configurations (40- and 42-residue peptide in a coil, helical, or beta-sheet form), interacting with all four HSV-1 glycoproteins. Our results confirm that specific forms of Aβ bind more strongly to HSV-1 glycoproteins than other forms, and that the most strongly-binding forms of Aβ interact with regions of HSV-1 glycoproteins that are crucial for host cell receptor binding. Overall, insights gained from these simulations may be used to guide the future design of novel drugs to prevent and/or treat HSV-1 infection in the brain. More broadly, this work provides a deeper understanding of the innate therapeutic properties of Aβ in the brain and may help to elucidate the true role of Aβ in other contexts like Alzheimer's disease.

References:

[1] Chen *et. al*, Amyloid beta: structure, biology and structure-based therapeutic development. *APS*, 38(9):1205-1235. (2017).

[2] Bourgade K *et. al*, Anti-Viral Properties of Amyloid-β Peptides. *Journal of Alzheimer's Disease Reports*, (6):599-606. (2022).

[3] Olubiyi *et. al*, Structures of the amyloid β-peptides Aβ1-40 and Aβ1-42 as influenced by pH and a d-peptide. *J. P. Chem. B,* 116(10):3280–3291. (2012).

197n Molecular-Level Insights into Hydrophilic Interaction Liquid Chromatography Via Molecular Simulations

Hsiao-Feng Liu, J. Ilja Siepmann, Mark R. Schure, Stephanie A. Schuster

Monte Carlo simulations are carried out to provide molecular-level insights into the retention mechanism in hydrophilic interaction liquid chromatography (HILIC). The stationary phases investigated here include a 9-nm cristobalite silica slit pore and a 9-nm amorphous silica slit pore. The mobile phases are water/acetonitrile mixtures at three compositions (neat acetonitrile, 5 and 10% v/v water in acetonitrile). The analytes include C1-C6 linear alkanes and alcohols. The simulations are performed at two temperatures, 303 and 333 K, and a pressure of 10 MPa. The simulations indicate that the stationary phase is retentive to the hydroxyl group (separating a primary alcohol from the linear alkane with the same number of carbon atoms) since the linear alcohols can form hydrogen bonds with the surface silanols and water molecules adsorbed on the silica surface. The addition of water reduces the retention increment of the OH group because of the stronger solvophilic effect. However, the contribution from the stationary phase to the retention of the OH group does not increase significantly as the water composition increases. Analyses of simulation trajectories provide insights into the structure of the HILIC phases, the orientational distribution of the solvent and analyte molecules, and the propensities of the silanol-analyte, silanol-water, and water-analyte hydrogen bond formation.

197o Li Ion Diffusion in Solid Electrolyte Analyzed Using Deep Generative Models.

Teppei Fukuya, Takayuki Nishio, Hiroya Nitta, Taku Ozawa, Kenji Yasuoka

Molecular dynamics (MD) is a useful simulation technique to evaluate materials properties, and wide varieties of materials are analyzed using both classical and ab-initio MD. Recently, high performance computing system enables us to compute systems consisting of even over millions of atoms utilizing parallel computation, however, long time simulation is still a problem since time evolution must be solved sequentially.

Machine learning technique, especially generative model, can be a surrogate model for the above MD simulation. Endo *et al.*, proposed a method called MD-GAN [1,2], which predicts time evolution of the (sub)system in an equilibrium state. We applied the method to predict the diffusivity of Li ions in solid state electrolyte. Set of trajectory data of the Li ions were obtained using ab-initio MD with SIESTA code [3]. The set of data were used as input data for MD-GAN, and predictive model was trained to predict long time behavior of the ions. The accuracy and effectiveness of the MD-GAN model are going to be discussed in this work.

1. Endo, Tomobe, and K. Yasuoka, *Proceedings of the AAAI Conference on Artificial Intelligence*, 32, 2192 (2018).
2. Kawada, et al., *Chem. Inf. Model*, 63, 76 (2023)
3. M. Soler *et al.*, *J. Phys. Condens. Matter*, 14, 2745 (2002).

197p Modeling Elastic Properties of Polyacrylamide Hydrogel Depending on Effective Structures

Seunghyok Rho, Sebin Kim, Won Bo Lee

Hydrogels have attracted significant interest in a range of applications due to their complex physical and chemical structures, which enable multiscale interactions. Polyacrylamide hydrogels, in particular, offer a wide range of mechanical properties and exhibit biocompatibility and versatile customization. However, simulating the macroscale properties of hydrogels is challenging due to the intricate multiscale interactions that occur within them. To optimize the effect of each scale while considering the trade-off relationship among them for numerical methods, it is crucial to understand how multiscale structures influence the mechanical properties of hydrogels. In this study, we present a coarse-grained molecular dynamics approach for predicting the elastic modulus of hydrogels based on effective chain lengths of defect-less network structures. We utilized a mixed force-field that accounted for both bonded and non-bonded interactions, with the bonded potentials being extracted from atomistic simulation results and the MARTINI coarse-grained force field used to simulate the non-bonded parts of the hydrogel. Our model utilized coarse-grained potentials to depict the initial deformation of uniaxial tensile, indicating Young's modulus, while also reflecting the effects of strain rates and swelling ratio on the modulus. Especially, strain rates should be treated with a caution due to rate differences of about 10 to the power of 6 to 10 per second, which could even be related to bond relaxation level inducing an exponential increase in modulus. Plus, the model was constructed with isotropic structures referencing experimental pre-gel conditions, being able to follow elastic properties' propensity depending on cross linker density.Our findings demonstrate the potential of using coarse-grained molecular dynamics methods to model and predict the elastic properties of hydrogels, despite the inherent randomness in their physical and chemical structures. Moreover, this approach could serve as a foundation for the development of various applications through versatile additives.

197q Using Deep Learning Potentials and Graph Lattice Models to Engineer Optimal Proton Conducting Membranes for Fuel Cells

Siddarth Achar, Leonardo Bernasconi, Karl Johnson

Development of new materials capable of conducting protons in the absence of water is crucial for improving the performance, reducing the cost, and extending the operating conditions for proton exchange membrane fuel cells. We present an atomistic simulation-based workflow to computationally design fuel cell membrane materials using deep learning potentials (DPs) and graph lattice models (GLMs). Our workflow was used to demonstrate that graphanol (hydroxylated graphane) conducts protons anhydrously with very low diffusion barriers. First, DPs were trained for graphanol that have near-density functional theory accuracy but require a very small fraction of the computational cost. We used these DPs to calculate proton self-diffusion coefficients as a function of temperature, to estimate the overall barrier of proton diffusion, and to characterize the impact of thermal fluctuations as a function of system size. The sensitivity of intrinsic energy barriers to the overall diffusion barrier for proton conduction was further assessed using large-scale GLMs. We discovered, using this workflow, the existence of transient hydrogen-bonded structures, which we indicate as *Grotthuss chains*, that determine the mechanism of proton conduction in graphanol. We showed that protons can rapidly hop along Grotthuss chains containing several hydroxyl groups aligned such that hydrogen bonds allow for conduction of protons forward and backward along the chain without hydroxyl group rotation. Long-range proton transport only takes place as new Grotthuss chains are formed by rotation of one or more hydroxyl groups in the chain. Thus, the overall diffusion barrier consists of a convolution of the intrinsic proton hopping barrier and the intrinsic hydroxyl rotation barrier. The results obtained from the application of our workflow also enable us to propose specific design rules for developing next-generation proton conducting materials with diffusion barriers lower than existing systems.

197r Exploring the Free Energy Landscape of Insulin Multimer Using Metadynamics

Yinhao Jia, Janani Sampath

Protein-protein interactions play a crucial role in the stability and function of proteins. In therapeutic applications, protein associations like the dimerization and hexamerization of insulin play an essential role during transport and storage. To provide insights into insulin's physiological role and to develop more effective strategies for optimizing its stability and efficacy, it is vital to delve deeper into the formation of higher-order structures and the associated dissociation processes. While experimental and computational approaches such as molecular dynamics simulations have been used to understand the aggregation process, atomistic-level details of insulin hexamerization remain elusive. However, it is challenging to study higher-order protein aggregation using classical MD simulations alone, due to the timescales involved. Enhanced sampling methods such as metadynamics are powerful techniques to overcome slow fluctuations in the system and effectively explore the underlying free energy landscape.

In this study, we investigate the process of insulin hexamerization by studying dimerization, followed by the association of three dimer pairs to form a hexamer. To effectively probe this process, we construct a new rotational collective variable (CV), which considers protein orientation and rotation relative to one another, offering detailed insights into the underlying association mechanisms. By employing well-tempered metadynamics (WT-MetaD) and parallel bias metadynamics (PB-MetaD), we probe the thermodynamics of insulin dimer and hexamer formation to understand association and dissociation processes. The established methodology and CV can be used as a tool for examining the thermodynamics of other protein systems, including their association free energy, augmenting our comprehension of intricate biological interactions.

197s Learning Analytical Bond-Order Parameters for Extrapolatable Neural Network Interatomic Potential of Multicomponent Materials

Minjoon Hong, Hoje Chun, Byungchan Han

Parametrization of neural network interatomic potentials has been widely studied to enable accelerated atomistic simulations at quantum chemical accuracy. The method proposed by Behler and Parrinello uses atom-centered symmetric functions (ACSF) to encode both geometric and chemical information, predicting atomic energy with high-dimensional neural networks. While this approach is powerful for fitting interatomic potentials to ab-initio calculations, limited extrapolation capability remains a significant challenge since the computational cost of data acquisition is notoriously high.

To address this issue, we introduce a regularization layer based on the analytical bond-order potential (BOP) model, imposing physical inductive bias to the model. Pairwise bond-order parameters are represented as functions of the local atomic environment using neural networks and are utilized in energy prediction with BOP. Our model demonstrates remarkable extrapolation capabilities for potential energy across unseen molecular geometries, which is unprecedented in conventional Behler-Parrinello neural network models.

Moreover, our model offers a physically meaningful interpretation of the learned potential energy by decomposing it into pairwise interactions. We validate the energy decomposition by comparing it with the quantum-chemically derived intrinsic bond strength index (IBSI), which shows excellent agreement. We demonstrate the pairwise interaction analysis to the path of the S_N2 reaction of methyl halides. It accurately captures the evolution of interatomic interactions throughout the reaction, showing the applicability of model on the study of chemically reactive systems.

197t Molecular Interactions, Architecture and Dynamics in Biomolecular Phase Separation

Gregory Dignon

Biomolecular condensates (BMCs) are non-stoichiometric assemblies of biomolecules which are essential for compartmentalization in biology, and generally form through a process of phase separation. They are promising as targets for drug development, as well as a unique strategy for various biotechnological applications. BMCs can be characterized by their materials properties, responses to external stimuli, and by multi-component phase diagrams. What underlies each of these properties is a complex network of molecular interactions between amino acid residues, nucleotides, specific interaction sites of proteins and nucleic acids, as well as solvent and co-solvent. By using multi-scale computational techniques, we study the interactions between molecules and subunits of molecules to understand how these interactions impact materials properties, and phase diagrams.

197u First-Principles Study on the Enhancement of Li-Ion Diffusion By Tunning the Anion Rotation in Various Solid-State Electrolyte

SuSeong Hyun, Byungchan Han

Solid-state electrolytes (SSE) are promising materials for use in lithium-ion batteries (LIB) due to their high energy density, wide temperature range, and reduced risk of leakage or explosion. Although high ionic conductivity is one of the most significant factors for electrolytes, it is still remaining a challenge for SSE. Intuitively, cation and anion doping are regarded as the most efficient way to boost Li-ion diffusion. Another way is to analyze the dynamics of cations and anions through the properties of the structure itself, such as anion rotation in plastic crystals. Here, we explore the mechanism of the anion rotation-assisted Li ion diffusion in various anion frameworks in solid-state electrolytes. The Sulfide framework shows the dynamic motion than the halide one, thereby incrementing the Li-ion diffusion. Also, polyanion frameworks such as $[PS_4]^{3-}$, play a key role in Li-ion mobility. We also investigate the collective motion between the anion and the Li-ion and construct a way to maintain its motion even at room temperature. This work suggests that the anion rotation may be a critical factor for enhancing ionic mobility and can be used as a useful guideline for designing a next-generation solid-state electrolyte.

197v Novel Sustainable Materials Design with Thermodynamics-Informed Machine Learning

Dinis Abranches, Edward Maginn, Yamil Colón

Scientists have traditionally employed trial-and-error methodologies to design novel materials, often complemented by basic heuristic rules or chemical intuition (e.g., "like dissolves like"). However, to date, this simplistic approach has led to the discovery and characterization of only a small fraction of all synthesizable compounds. This limitation is aggravated when studying mixtures, particularly those that show potential in minimizing the environmental impact of the chemical industry, as the number of all possible molecule combinations is virtually infinite. The massive amount of resources and extreme cost necessary to explore this chemical landscape is simply too demanding to be feasible.

Data-driven methods such as machine learning are promising alternatives to classical trial-and-error methodologies. The premise of these approaches consists of training machine learning models using existing datasets of known materials, and subsequently applying them to guide the development of new compounds by predicting physicochemical properties relevant to any given target application. However, a severe limitation of this approach lies on the representation of molecules in a machine-intelligible way (i.e., converting molecular structures to sets of numerical values that can be understood by mathematical machine learning models). In fact, most molecular representations proposed in the past lead to overly complex models that require a tremendous volume of experimental data to be properly trained.

To bridge the gap between small, scarce datasets and data-driven approaches, this work aims at designing novel sustainable materials using thermodynamics-informed machine learning models, with a particular focus on green solvents. The use of thermodynamics information was achieved in two separate ways. The first was to use Gaussian processes (GPs) and active learning (AL) to describe activity coefficients (a single property), which can then be combined with thermodynamic models to predict a plethora of different solvent-related properties (e.g., phase equilibria). Relying on synthetic data generated from an excess Gibbs energy model, GPs were found to accurately describe the activity coefficients of several binary mixtures across large composition and temperature ranges. Moreover, GPs could estimate their own uncertainty and identify composition/temperature regions where activity coefficient data provides the most information to the models. This was leveraged to build AL algorithms targeted at modelling phase equilibria. In many cases, a single active-learning-acquired data point was sufficient to describe the phase diagrams studied. The ability of AL to greatly reduce the amount of data needed to obtain accurate models was further verified on experimental case studies, namely individual ion activity coefficients, the solid-liquid and vapor-liquid equilibrium of deep eutectic solvents, and phase equilibria in ternary mixtures.

The second thermodynamics-informed approach was to develop a molecular descriptor that encodes thermodynamic information, namely molecular polarity, and that possesses a small, fixed size. In this sense, sigma profiles, a type of molecular descriptor obtained through quantum chemistry calculations, were used to train convolutional neural networks (CNNs) that accurately correlate and predict a wide range of physicochemical properties (molar masses, normal boiling temperatures, vapor pressures, densities, refractive indexes, and aqueous solubilities). To boost their performance, the architecture and hyperparameters of each CNN were optimized using a battery of algorithms, particularly Bayesian Optimization and Local Search. Furthermore, it was shown that thermodynamic conditions, namely temperature, can also be used as additional inputs to broaden the applicability of the models.

Finally, inverse design workflows based on this type of thermodynamics information (active learning on the sigma profile space) were also constructed, which enable the computational design of tailor-made materials bounded by a set of physicochemical properties selected based on a given application of interest. The results of this work constitute a major breakthrough in the way novel materials can be designed, as the machine learning models developed are robust, accurate, and generic enough to be applied across several families of compounds and materials.

197w Molecular Dynamics Simulations and Graph-Theoretic Analyses of Nanostructure Formation in Ionic Liquids

Lisa Je, Victor M. Zavala, Reid Van Lehn

Ionic liquids (ILs), such as 1-ethyl-3-methylimidazolium tetrafluoroborate (EMIM-BF_4), have been scrutinized as viable electrolytes for electrochemical reactions due to their highly tunable cation/anion properties, intrinsic conductivity, wide electrochemical stability window, low volatility, and strong electrostatic interactions [1-3]. However, the molecular structure of ILs in bulk and dilute solutions remain poorly understood and prevents the rational design of ILs [4-6]. Molecular simulations can address this problem by modeling the formation of IL nanostructures that is difficult to observe via experiments. Using both the concentration of IL-solvent mixtures knobs, we performed atomistic molecular dynamics (MD) simulations to quantify concentration-dependent changes in IL structure. From MD trajectories, we determined how interactions between the IL and solvent impact structure through graph-theoretic representations of IL nanostructure. We show that EMIM-BF_4 exhibits variations in the tendency to form both charged and neutral molecular-scale clusters as a function of concentration ranging from the dilute to neat regimes. Via these graph-theoretic methods, we investigated quantitative parameters of IL structure (such as length of nonpolar alkyl group) that influence IL structural clustering such as: size of clusters, number of clusters, and number of neutral clusters. These quantitative metrics will permit understanding of how IL structure impacts electrolyte properties which is necessary for fine-tuning electrochemical systems.

[1] K. Shimizu, C. E. S. Bernardes, and J. N. Canongia Lopes, 'Structure and Aggregation in the 1-Alkyl-3-Methylimidazolium Bis(trifluoromethylsulfonyl)imide Ionic Liquid Homologous Series,' *The Journal of Physical Chemistry B,* vol. 118, no. 2, pp. 567-576, 2014.

[2] K. Motobayashi, Y. Maeno, and K. Ikeda, 'In Situ Spectroscopic Characterization of an Intermediate of CO2 Electroreduction on a Au Electrode in Room-Temperature Ionic Liquids,' *The Journal of Physical Chemistry C,* vol. 126, no. 29, pp. 11981-11986, 2022.

[3] B. A. Rosen *et al.*, 'In Situ Spectroscopic Examination of a Low Overpotential Pathway for Carbon Dioxide Conversion to Carbon Monoxide,' *The Journal of Physical Chemistry C,* vol. 116, no. 29, pp. 15307-15312, 2012.

[4] B. A. Rosen *et al.*, 'Ionic Liquid-Mediated Selective Conversion of CO2 to CO at Low Overpotentials,' (in English), *Science,* Article vol. 334, no. 6056, pp. 643-644, Nov 2011, doi: 10.1126/science.1209786.

[5] M. A. Gebbie *et al.*, 'Long range electrostatic forces in ionic liquids,' *Chemical Communications,* vol. 53, no. 7, pp. 1214-1224, 2017.

[6] B. Liu, W. Guo, and M. A. Gebbie, 'Tuning Ionic Screening To Accelerate Electrochemical CO2 Reduction in Ionic Liquid Electrolytes,' *ACS Catalysis,* vol. 12, no. 15, pp. 9706-9716, 2022.

[7] A. M. Smith, A. A. Lee, and S. Perkin, 'The Electrostatic Screening Length in Concentrated Electrolytes Increases with Concentration,' *The Journal of Physical Chemistry Letters,* vol. 7, no. 12, pp. 2157-2163, 2016.

197x How Do Self-Interaction Errors Associated with Stretched Bonds Affect Barrier Height Predictions?

Priyanka Bholanath Shukla, Prakash Mishra, Tunna Baruah, Rajendra Zope, Koblar Jackson, Karl Johnson

Density functional theory suffers from self-interaction errors that generally result in the underestimation of chemical reaction barrier heights. This is commonly attributed to the tendency of density functional approximations to overstabilize delocalized densities that typically occur in the stretched bonds of transition state structures. The Perdew–Zunger self-interaction correction (PZSIC) and locally scaled self-interaction correction (LSIC) improve the prediction of barrier heights of chemical reactions, with LSIC giving better accuracy than PZSIC on average. These methods employ an orbital-by-orbital correction scheme to remove the one-electron SIE.

In the context of barrier heights, this allows an analysis of how the self-interaction correction (SIC) for each orbital contributes to the calculated barriers using Fermi–Löwdin orbitals. We hypothesize that the SIC contribution to the reaction barrier comes mainly from a limited number of orbitals that are directly involved in bond-breaking and bond-making in the reaction transition state. We call these participant orbitals (POs), in contrast to spectator orbitals (SOs) which are not directly involved in changes to the bonding.

We test this hypothesis for the reaction barriers of the BH76 benchmark set of reactions. We find that the stretched-bond orbitals indeed make the largest individual SIC contributions to the barriers. These contributions increase the barrier heights relative to LSDA, which underpredicts the barrier. However, the full stretched-bond hypothesis does not hold in all cases for either PZSIC or LSIC. There are many cases where the total SIC contribution from the SOs is significant and cannot be ignored. The size of the SIC contribution to the barrier height is a key indicator. A large SIC correction is correlated to a large LSDA error in the barrier, showing that PZSIC properly gives larger corrections when corrections are needed most. A comparison of the performance of PZSIC and LSIC shows that the two methods have similar accuracy for reactions with large LSDA errors, but LSIC is clearly better for reactions with small errors. We trace this to an improved description of reaction energies in LSIC.

197y Pysages: Enhanced Sampling for Ab Initio Dynamics and Machine Learning Potentials

Gustavo Perez Lemus, Pablo Zubieta, Yezhi Jin, Yinan Xu, Juan J. de Pablo

PySAGES is a user-friendly Python library that enables enhanced sampling in molecular dynamics simulations by providing a simple interface to write and leverage complex collective variables and enhanced sampling methods. In this work, we demonstrate the coupling of PySAGES with Ab Initio integrators using the ASE interface, which facilitates enhanced sampling in systems that require first-principles accuracy. Additionally, we showcase the use of PySAGES as a tool for studying the robustness of machine-learned force fields by evaluating different examples of DeePMD, GAP, and Graph Neural Network potentials and assessing their reliability in preserving the behavior of a system in terms of selected molecular descriptors.

Our work highlights the versatility of PySAGES and its potential to accelerate the study and analysis of molecular dynamics simulations, as well as the design of molecules and materials in various fields.

197z Permeation of Glycoside Molecules through Lipid Bilayers

Jinhui Li, Viviana Monje-Galvan

Saponins are glycosides ubiquitously distributed in the plant kingdom. They have received extensive attention for their potential medicinal value. Recent studies showed that their biological activities are closely modulated by interaction with lipid bilayers. For example, membrane permeation, can directly affect the passive diffusion of saponins, which is also the dominant pathway for saponins to pass through the intestinal tract, blood vessel wall, or other human tissues. Undoubtedly, the absorption of saponin will be directly related to its medicinal efficacy, and clarification of this permeation process is of great interest to the pharmaceutical industry. Here, we combined molecular dynamics simulations with transition-tempered metadynamics to study the permeation process of a monosaccharide saponin (SAM). The time-structure based independent component analysis (tICA) was also applied to design collective variables to accelerate the convergence of free energy profile calculation. Simulations allowed us to characterize the permeation details of saponin through lipid bilayers, which primarily occurs by both hydrophilic interactions of SAM sugar group with lipid headgroups and hydrophobic interactions between SAM aglycone ring and lipid tails. We tentatively proposed the following general permeation mechanism: initially, due to the hydrophilic interaction between the SAM sugar group and lipid headgroups, the molecule approaches the bilayer surface. Then, SAM flips into the bilayer core because of the hydrophobic interaction of SAM aglycone with lipid tails. As the simulation progresses, SAM gradually moves down to the other leaflet and flips again such that its sugar group points towards the lipid/water interface. Finally, through interaction with water, SAM is pulled out of the bilayer. During the permeation process, the lipid bilayer deforms, and water molecules penetrate into hydrophobic core as the SAM flips between leaflets. We carry simulations with different membrane models and found that lipid species modulate the permeation mechanism, resulting in different free energy profiles. In summary, this work provides insights about the permeation process of saponin through lipid bilayers at the atomistic level. Understanding the passive diffusion mechanism of small molecules can also contribute to decrease the cost and time of developing novel antibiotic and anticancer treatments.

197aa Machine Learning Anisotropic Coarse-Grained Potentials

Marjan Albooyeh, Eric Jankowski

Molecular dynamics (MD) simulations enable prediction of equilibrium structure and properties of complex molecular systems, but slow relaxation times and high computational cost of highly-resolved (all-atom and united atom) representations prohibits the interrogation of many multi-million-atom morphologies. This is where coarse-grained molecular dynamics (CGMD) proves useful, as it reduces the computational complexity by grouping atoms into larger simulation elements. Here we focus on the use of non-spherical beads to more accurately capture the structure of materials with complex shapes and directional dependencies that are difficult to represent with spherical simulation elements.

We combine machine learning (ML) with CGMD to learn the forces and torques between nonspherical beads that will result in equivalent trajectories to all-atom simulations. A proof-of-concept neural network model is developed using PyTorch for the organic semiconductor PPS, which takes in the relative positions and orientations of atomistic simulations to learn a latent representation of the local pair interactions by incorporating neighbor data. The trained neural network is used to infer the forces and torques acting on coarse-grained simulation elements. We compare the accuracy of models trained against data sets of varying size, and those biased to include configurations where the intermolecular potential gradients are largest. We implement the resulting ML potentials in the HOOMD-Blue simulation engine and compare against all-atom rigid models of PPS quantitatively and qualitatively. We conclude with a discussion of future directions and network architectures that may offer further performance and accuracy enhancements. In sum we develop tools for performing efficient large-scale structural predictions of complex molecules including organic semiconductors, semiconducting polymers, and enable scientific interrogations related to these materials.

This material is based upon work supported by the National Science Foundation under Grant No. 1653954. This research was partially supported by the National Aeronautics and Space Administration (NASA) under the University Leadership Initiative program; grant number 80NSSC20M0165.

197ab Modeling Electrode-Electrolyte Interfacial Effects during Specific Alkali Metal Cation Adsorption Using a DFT/FF-MD Approach

Andrew Wong, Bolton Tran, Scott T. Milner, Michael Janik

Significant efforts have been focused on developing active and selective electrocatalysts to facilitate electrochemical transformations in batteries, electrolysis cells, fuel cells, and carbon capture technologies. It has been known the altering the identity of the ions in the electrolyte has significant effects on the rates and selective on electrocatalytic reactions such as CO_2 reduction, HER, OER, N_2 reduction, and water oxidation. Recently, Waegele et al. was able to demonstrate that specifically adsorbed alkali metal cations on Au electrodes not only alter the rate of CO_2 to CO but is sensitive to the identity of the cation and ability to shed its hydration shell.[1] Modeling specifically adsorbed cations and its interactions within the electrochemical double layer (EDL) is complex, requiring careful modeling choices on how solvation and electrification within this system to be made. Our prior work for NH to NH_2 reduction on Rh (111) showed that predicted barriers for this system can vary by 1 eV based on the model parameters used to describe the EDL. In this work, we used a DFT/FF-MD approach to investigate the thermodynamic equilibrium of alkali metal cation adsorption across late transition metal surfaces while identifying the sensitivity made from the choices within our model. We used DFT methods to study the interactions between the cation and electrode and nearby explicit static H_2O, representing micro-solvation from the first hydration shell of the adsorbed cation. We utilized a "post-DFT" Helmholtz model to correct the DFT simulation to constant potential and incorporate electrification using two discernable parameters, the dielectric constant and width of the EDL. Force-Field Molecular Dynamics (FF-MD) are incorporated to estimate the dynamic effects of solvation upon cation adsorption, which also informs our Helmholtz model suitable dielectric constants and EDL widths to use.

Thermodynamic equilibrium of specific cation adsorption across different late transition metal surfaces are studied by predicting the equilibrium potential (U^0), the applied potential of the system where cation adsorption is at equilibrium. We observed that predicted U^0's for alkali metal cations can vary over 1 V-NHE based on the model used to predicted solvation (explicit static H_2O, continuum solvation, and FF-MD). For incorporating the interactions of electrification, U^0's are sensitive to the choice of the width and dielectric constant of the EDL, showing variations in U^0 of 3 V-NHE span for dielectric constants from 2 to 8. Magnitude of the EDL effects can be predicted from the magnitude of the dipole moment change upon adsorption. Lastly, we demonstrate that our DFT/FF-MD approach can rationally tune the model parameter to agree with potential ranges observed on Au (111) from Waegele et al.

References:

1. Ovalle et al. Nature Catalysis. Vol 5. July 2022. 624-632.

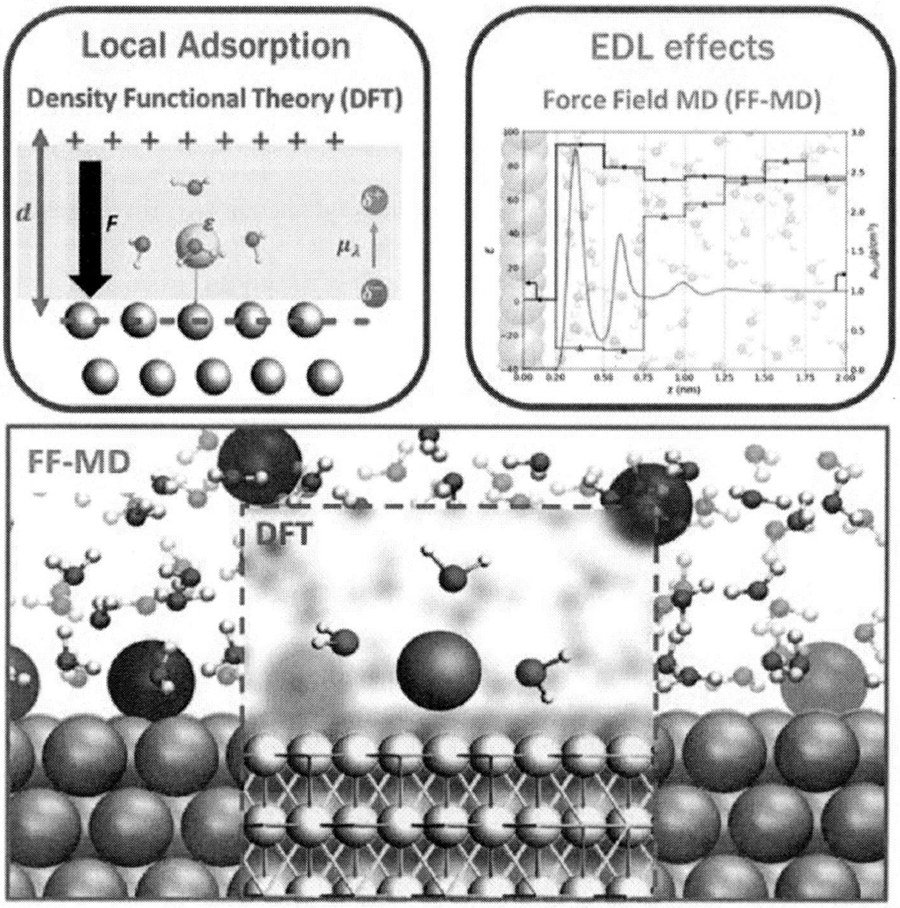
Local Adsorption
Density Functional Theory (DFT)
EDL effects
Force Field MD (FF-MD)
FF-MD
DFT

197ac Reproducible Workflows for Parameterizing and Simulating Models of Complex Conjugated Copolymers for Organic Photovoltaics

Madilyn Paul, Eric Jankowski

Organic semiconductors that can self-assemble into morphologies with high charge mobility have the potential to transform green energy generation, but measuring or predicting the stable morphologies and their charge mobilities is difficult. Here we develop computational workflows for screening the morphologies of organic photovoltaic (OPV) macromolecules and copolymers. We employ the Espaloma[1] toolkit to parameterize all-atom forcefields, signac-flow[2] to manage simulation workspaces, MoSDeF tools[3] to initialize simulation volumes, and HOOMD-Blue[4] to perform molecular dynamics simulations. We compare predicted morphologies between all-atom and united atom representations with a variety of methods of handling partial charges for a set of complex donor-acceptor copolymers[5] and justify the use of the coarser models to predict morphology in these systems. Finally, we summarize short-range structural motifs and long-range periodicities that emerge across conjugated polymers with varying backbone rigidity.

This material is based upon work supported by the National Science Foundation under Grant No. 1653954.

(1) Wang, Y.; Fass, J.; Kaminow, B.; Herr, J. E.; Rufa, D.; Zhang, I.; Pulido, I.; Henry, M.; Chodera, J. D. End-to-End Differentiable Molecular Mechanics Force Field Construction. *Chem. Sci.* **2022**, *13* (41), 12016–12033. https://doi.org/10.1039/D2SC02739A.

(2) Adorf, C. S.; Dodd, P. M.; Ramasubramani, V.; Glotzer, S. C. Simple Data and Workflow Management with the Signac Framework. *Comput. Mater. Sci.* **2018**, *146*, 220–229. https://doi.org/10.1016/j.commatsci.2018.01.035.

(3) Cummings, P. T.; MCabe, C.; Iacovella, C. R.; Ledeczi, A.; Jankowski, E.; Jayaraman, A.; Palmer, J. C.; Maginn, E. J.; Glotzer, S. C.; Anderson, J. A.; Ilja Siepmann, J.; Potoff, J.; Matsumoto, R. A.; Gilmer, J. B.; DeFever, R. S.; Singh, R.; Crawford, B. Open-source Molecular Modeling Software in Chemical Engineering Focusing on the Molecular Simulation Design Framework. *AIChE J.* **2021**, *67* (3). https://doi.org/10.1002/aic.17206.

(4) Anderson, J. A.; Glaser, J.; Glotzer, S. C. HOOMD-Blue: A Python Package for High-Performance Molecular Dynamics and Hard Particle Monte Carlo Simulations. *Comput. Mater. Sci.* **2020**, *173*, 109363. https://doi.org/10.1016/j.commatsci.2019.109363.

(5) Danielsen, S. P. O.; Bridges, C. R.; Segalman, R. A. Chain Stiffness of Donor–Acceptor Conjugated Polymers in Solution. *Macromolecules* **2022**, *55* (2), 437–449. https://doi.org/10.1021/acs.macromol.1c02229.

197ad Computational Screening and Designing of Solid Materials for CO_2 Capture Technology

Yuhua Duan

Capture of CO_2 is a critical component in producing valuable reuse products from fossil fuel-based processes. It is generally accepted that current commercial technologies for capturing CO_2 are too energy intensive and thus cost prohibitive for implementation on coal based power plants. Hence, there is a critical need for development of new materials that can capture CO_2 reversibly with acceptable energy and cost performance for these applications. Accordingly, solid sorbents have been reported in several previous studies to be promising candidates for CO_2 sorbent applications through a reversible chemical transformation due to their high CO_2 absorption capacities at moderate working temperatures. By combining thermodynamic database mining with first principles density functional theory and phonon lattice dynamics calculations, a theoretical screening methodology to identify the most promising CO_2 sorbent candidates from the vast array of possible solid materials have been proposed and validated at the National Energy Technology Laboratory (NETL). The advantage of this method is that it identifies the thermodynamic properties of the CO_2 capture reaction as a function of temperature and pressure without any experimental input beyond crystallographic structural information of the solid phases involved. The calculated thermodynamic properties of different classes of solid materials versus temperature and pressure changes were further used to evaluate the equilibrium properties for the CO_2 adsorption/desorption cycles. Per the requirements imposed by the pre- and post- combustion technologies and based on our calculated thermodynamic properties for the CO_2 capture reactions by the solids of interest, we are able to identify only those solid materials for which lower capture energy costs are expected at the desired pressure and temperature conditions. These CO_2 sorbent candidates were further considered for experimental validations.

However, at a given CO_2 pressure, the turnover temperature (T_t) of an individual solid capture CO_2 reaction is fixed and may be outside the operating temperature range (ΔT_o) for a particularly capture technology. In order to shift such T_t for a solid into the range of ΔT_o, its corresponding thermodynamic property must be changed by changing its molecular structure by mixing with other materials or doping with other elements. Our results demonstrate that by mixing or doping two or more materials to form a new material, it is possible to synthesize new CO_2 sorbents which can fit the needs for industrial applications.

Recent experimental and calculated results show that some solid sorbents can serve as bi-functional materials: CO_2 sorbent and CO oxidation catalyst. Such dual functionality could be used for removing both CO and CO_2 after water-gas-shift to obtain H_2 with high purity.

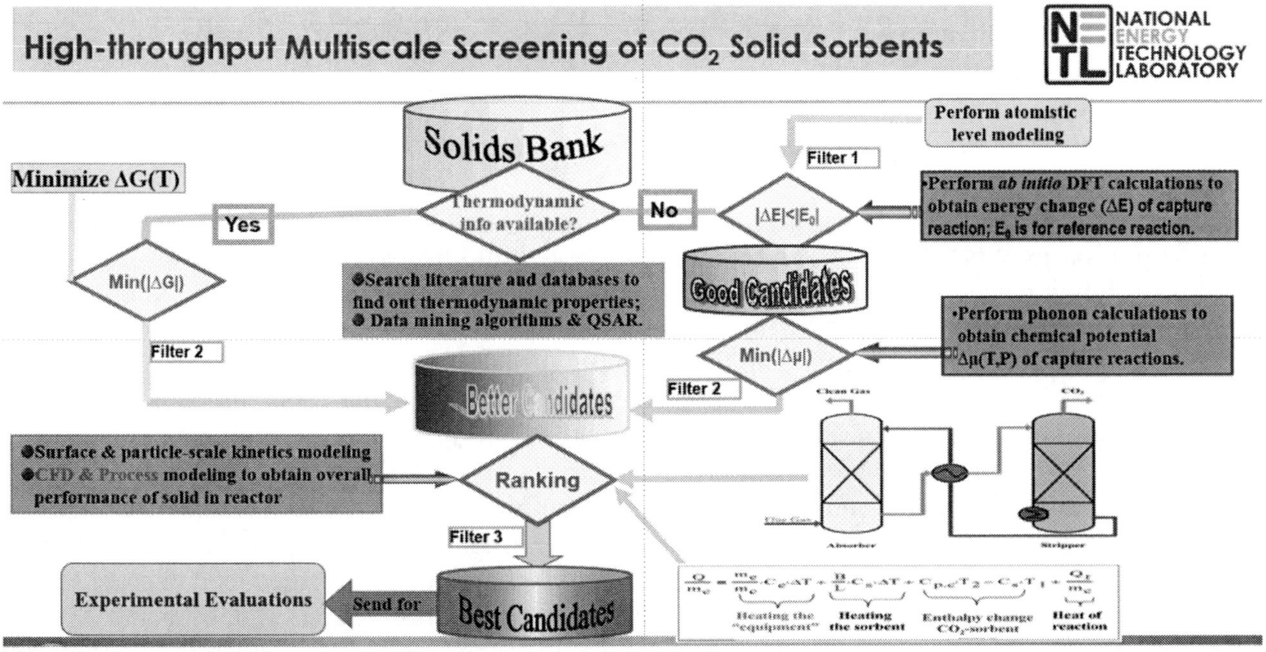

High-throughput Multiscale Screening of CO2 Solid Sorbents
NATIONAL ENERGY TECHNOLOGY LABORATORY
Solids Bank
Minimize ΔG(T)
Thermodynamic info available?
Yes
No
Min(|ΔG|)
Search literature and databases to find out thermodynamic properties;
Data mining algorithms & QSAR.
Perform atomistic level modeling
Filter 1
|ΔE|<|E0|
Perform ab initio DFT calculations to obtain energy change (ΔE) of capture reaction; E0 is for reference reaction.
Good Candidates
Perform phonon calculations to obtain chemical potential Δμ(T,P) of capture reactions.
Min(|Δμ|)
Filter 2
Filter 2
Better Candidates
Surface & particle-scale kinetics modeling
CFD & Process modeling to obtain overall performance of solid in reactor
Ranking
Filter 3
Experimental Evaluations
Send for
Best Candidates
Clean Gas
CO2
Absorber
Stripper
Heating the "equipment"
Heating the sorbent
Enthalpy change CO2-sorbent
Heat of reaction

197ae Probing the Energy Landscape and Hierarchical Self-Assembly of Magnetic Handshake Panels

Andreia L. Fenley, Chrisy Xiyu Du, Ran Niu, Itai Cohen, Michael P. Brenner, Paul L. McEuen, Julia Dshemuchadse

In the field of programmable self-assembly, where simple building blocks can form highly complex structures following specific local rules of interaction, the desired criteria for these interactions are that they be strong, to enable stable final configurations; specific, to allow for control over which units interact; and long-range, to speed up the assembly process. Embedding different patterns of magnetic dipoles into rigid panels tackles the problem of short-range and weak interactions. The patterning lends specificity to these "magnetic handshake" building blocks, while the magnetic dipoles impart strong, long-range interactions [1]. When assembled, the panels organize hierarchically, first into chains, and then those chains combine to form dense stacks. Differences in the panel types govern the degree of condensation and the order of the observed phase transitions, as well as the final morphologies, allowing us to characterize which panel features lead to structural differences at a larger length scale. We seek to build upon those observations and, by better understanding these kinetic pathways, to obtain control over the emergent structures, as well as the path that the system takes to assemble them. This work will elucidate both kinetic and thermodynamic states of the system, as well as the associated pathways, allowing for the spatiotemporal manipulation of the simple panels into their varied and complex self-assembled final structures.

[1] Ran Niu, Chrisy Xiyu Du, Edward Esposito, Jakin Ng, Michael P. Brenner, Paul L. McEuen, Itai Cohen, 'Magnetic handshake materials as a scale-invariant platform for programmed self-assembly', Proc. Natl. Acad. Sci. 116, 24402–24407 (2019).

197af Data-Driven Simulation of Chromatin Structure and the Epigenetic Code

Soren Kyhl, Aria Coraor, Juan J. de Pablo

The three-dimensional structure of DNA in the cell nucleus is increasingly recognized as an important factor in disease and development. Interestingly, chromatin organization has been correlated with a number of chemical modifications deposited 'on top of' the genome (i.e. epi-), spawning the field of epigenetics. These epigenetic modifications do not alter any base pairs, but nevertheless constitute a one-dimensional code which has been hypothesized to template and regulate the three-dimensional folding of DNA. Such patterns of epigenetic marks and DNA structure are highly specific to cell type, organism development, and even disease state. Impressively, multiple experimental techniques have arisen to measure aspects of chromatin structure, such as pairwise contact frequencies. However, no experimental technique furnishes the complete three-dimensional structure. This is further complicated by the fact that chromatin exists in a statistical ensemble of structures, with high variability from cell to cell. We present results using techniques from statistical physics to reconstruct the most likely ensemble of chromatin states consistent with the experimental data. This workflow complements experiments in order to elucidate the structure of chromosomes at increased resolution. With these models, we investigate the sequence-to-structure relationship between epigenetic marks and chromosome structure and ask the question: do epigenetic marks alone carry sufficient information to template three-dimensional chromosome organization?

197ag Computationally Exploring Structure-Property Relationships of Thermal Transport in Metal-Organic Frameworks Using High-Throughput Screening and Machine Learning

Meiirbek Islamov, Hasan Babaei, Christopher Wilmer

Metal-Organic Frameworks (MOFs) are emerging as a highly promising class of materials for applications such as gas storage, separation, and catalysis due to their large surface area, tunable pore geometry, and high porosity. However, the practical utilization of MOFs is often limited by the thermal energy generated during the exothermic adsorption process. Unfortunately, the thermal transport properties of MOFs have not been studied extensively, resulting in a lack of fundamental knowledge about the structure-thermal conductivity relationships necessary for designing MOFs with specific thermal properties. To fill this knowledge gap, we conducted the first computational high-throughput screening of hypothetical MOFs using classical molecular dynamics simulations and the Green-Kubo method. We screened over 10,000 hypothetical MOFs generated by the ToBaCCo-3.0 code, accounting for significant structural and compositional features, including pore size, density, surface area, node-linker mass mismatch, and metal node connectivity. Our analysis revealed that small pores, high density, small node-linker mass mismatch, and four-connected metal clusters are associated with high thermal conductivity.

Furthermore, we trained a graph convolutional neural network on the thermal conductivities of nearly 10,000 hypothetical MOFs. This machine learning model can predict MOF thermal conductivity with a mean absolute error of 0.05 W m^{-1} K^{-1}, which provides an efficient means of exploring the vast design space of MOFs. Additionally, we examined the effect of randomly distributed missing linker and missing cluster defects on the thermal conductivity of UiO-66 and HKUST-1, two well-known MOFs. We also investigated the thermal conductivity of three experimentally determined correlated defect nanodomains of UiO-66 with underlying topologies of bcu, reo, and scu nets. Our findings suggest that both randomly introduced missing linker and missing cluster defects reduce thermal conductivity, while the correlated missing linker defect nanodomain (bcu topology) displayed increased thermal conductivity than pristine UiO-66. Harmonic lattice dynamics calculations also supported these results, indicating an increase in phonon group velocity. In conclusion, our study provides crucial new insights into the design principles of MOFs with tailored thermal properties and highlights the significance of considering structural characteristics and defects when designing high-performance MOFs for various applications. Our application of classical molecular dynamics simulations, machine learning, and defect analysis offers a comprehensive framework for understanding and controlling the thermal transport properties of MOFs.

197ah Common Framework Mutations Impact Antibody Interfacial Dynamics and Flexibility

Emily Rhodes, Jonathan Faris, Brian Petersen, Kayla G. Sprenger, Emma C. Aldrich

With the flood of engineered antibodies, there is a heightened need to elucidate the structural features of antibodies that contribute to specificity, stability, and breadth. While antibody flexibility and interface angle have begun to be explored, design rules have yet to emerge, as their impact on the metrics above remains unclear. Furthermore, the purpose of framework mutations in mature antibodies is highly convoluted. To this end, a case study utilizing molecular dynamics (MD) simulations was undertaken to determine the impact framework mutations have on the V_H-V_L interface. We further sought to elucidate the governing mechanisms by which changes in the V_H-V_L interface angle impact structural elements of mature antibodies by looking at root mean squared deviations (RMSD), V_H-V_L interface angle, root mean squared fluctuations (RMSF) and solvent exposed surface area (SASA). Overall, our results suggest framework mutations can significantly shift the distribution of V_H-V_L interface angles which leads to increased flexibility by changing which portions of the antibody have a greater solvent exposed surface area. The data highlights the need to reject the dogma of static antibody crystal structures and exemplifies the dynamic nature of these proteins in solution. Findings from this work further demonstrate the importance of framework mutations on antibody structure and lay the foundation for establishing design principles to create antibodies with increased specificity, stability, and breadth.

197ai Effect of Molecular Structure on the HOMO-LUMO GAP Using Atomic Signatures As Molecular Descriptors

Donald Visco Jr., Ahmed Mohamed, Karl Breimaier, David Bastidas

Compounds with low HOMO–LUMO gap E_{gap} are in need as they are kinetically unstable making them reactive and desirable for different applications. However, the search for new organic compounds with a low E_{gap} is an expensive endeavor due to the exponentially increasing pool of virtual compounds. Thus, there is a need to construct a robust, predictive, and computationally inexpensive quantitate structure property relationship (QSPR) to identify and screen for structures possessing low E_{gap}. Herein, a QSPR is proposed that utilizes atomic Signatures as molecular descriptors correlating the molecular structure with the E_{gap}. A dataset consisting of 112 organic compounds was used to construct the QSPR using forward stepping multilinear regression (FSMLR) with leave-one-out cross validation (LOOCV), achieving a regression coefficient (r^2) and predictability (q^2) of 0.86 and 0.76, respectively. The QSPR was able to infer relations between different structural motifs (i.e. atomic Signatures) and the E_{gap}. It was found that the atomic Signature encapsulating aromatic bonds, [C](p[C]p[C][H]), was able to explain around 50% of the variance in the data, showing its significance in affecting the property of interest. The five most significant Signatures were found to negatively affect the E_{gap}, due to them encapsulating π–bonds, heteroatoms, and aromatics. This is attributed to high electron density, making these structural motifs reactive sites, specifically due to $\pi - \pi$ interactions, lone pair electrons, and/or delocalized π–bonds. In constant the presence of terminal nitrogen or internal alkane chain groups causes an increase in the E_{gap}, which can be attributed to steric hindrance and non–conjugated bonding. Finally, an external test set were used to further elucidate the predictive ability of the model – an r^2 of 0.91 was achieved. Overall, the constructed QSPR can be used as a reliable initial screening tool to identify potential candidates possessing low E_{gap} values

197aj Exploring the Factors behind Adsorbate-Adsorbate Interactions

Shyama Charan Mandal, Frank Abild-Pedersen

In heterogeneous catalysis, adsorbate-adsorbate interaction can greatly affect reaction rates. Understanding such interactions are crucial for determining intrinsic kinetics and catalytic activity. Herein, we present a model for the interaction among small adsorbates such as C, H, and O on the Pt(111) surface. This model considers the modification of the metal surface and the adsorption strength resulting from other adsorbates in the vicinity of the adsorption site. For the same adsorbates, the binding energy of the metal atoms of the active site decreases as the number of C, H, or O adsorbates in their surroundings on the Pt(111) surface increases. Specifically, the binding energy decreases from -6.4371 to -5.4003 eV for C, from -6.4371 to -6.3351 eV for H, and from -6.4371 to -5.2154 eV for O. It is noteworthy that the presence of other adsorbates can significantly affect the behavior of a specific adsorbate, these effects cannot be explained solely by the electronic structure. The electrostatic interactions among various adsorbates also play a crucial role. Remarkably, the changes in the predicted and calculated C, H, and O adsorption energy on the Pt(111) surface in the presence of other adsorbates exhibit a linear scaling behavior with the number of adsorbates in the surroundings, indicating that the electrostatic interactions among various adsorbates depend on the type and number of interacting adsorbates. When the same type of adsorbates are present, such as C-C, O-O, and H-H, the repulsive nature between them is greater than that between different types of adsorbates, such as C-O, C-H, and O-H. The influence of the electronic and electrostatic components on the adsorbate-adsorbate interaction is more pronounced for C and O, while it is less significant for H. The electronic and electrostatic effects can be determined by simple scaling with number of adsorbates, and the metal binding energies can be utilized to predict the adsorbate binding energies (Figure 1). By considering the electronic and electrostatic components and the presence of different types of adsorbates, the model offers a significant advantage of predicting the adsorbate-adsorbate interaction. Ultimately, this can be achieved without the need for extensive computational calculations, making it a valuable tool for both theoreticians and experimentalists in the field of heterogeneous catalysis.

Figure 1: The binding energy of metal (ΔE_M), predicted adsorbate binding energy considering changes in the electronic structure and calculated adsorbate binding energy as a function of number of adsorbates present in the surrounding.

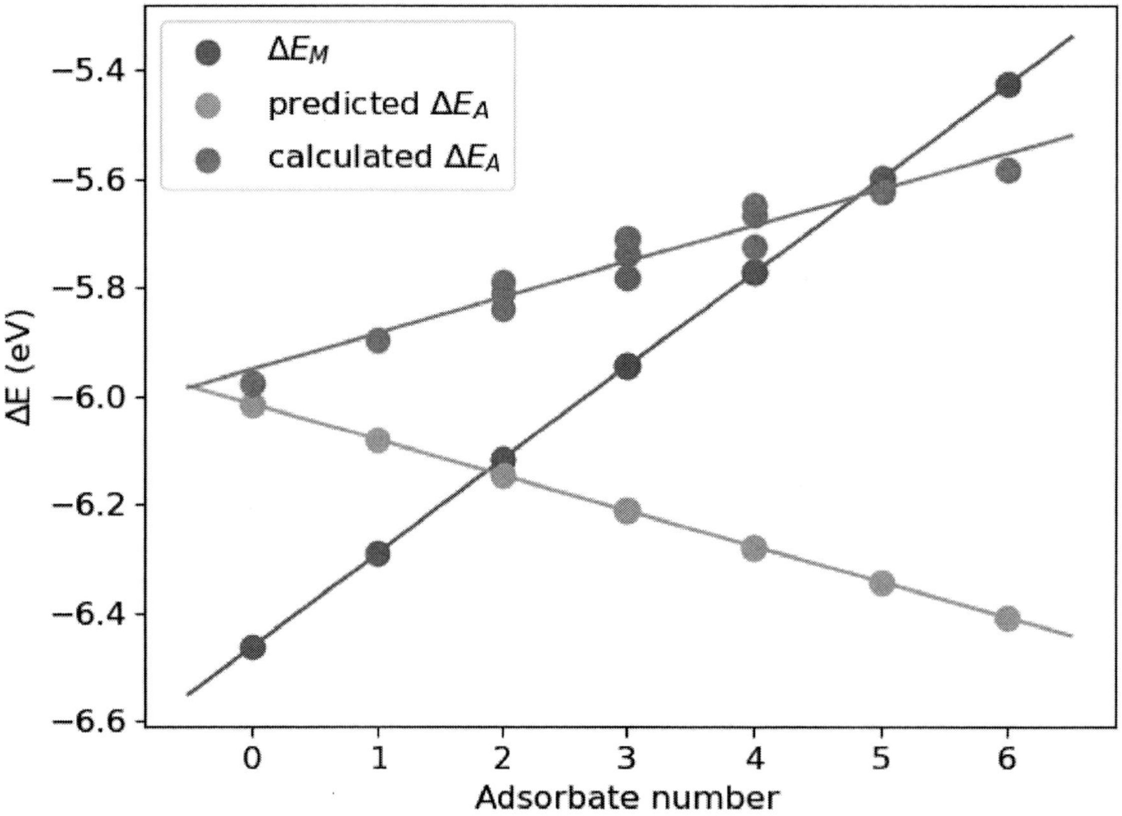

ΔE_M
predicted ΔE_A
calculated ΔE_A
−5.4
−5.6
−5.8
−6.0
−6.2
−6.4
−6.6
ΔE (eV)
0
1
2
3
4
5
6
Adsorbate number

197ak Towards the Development of Explainable Graph Neural Network Model for Predicting Viral Antibody-Antigen Binding Interfaces

Emma Stevens, Subhashis Hazarika, Srirupa Chakraborty

The phenomenon of antigen epitopes being specifically recognized by antibodies is not only a critical aspect of our immune response, but is also imperative for vaccine and therapeutic-antibody design. Consequently, detailed characterization and prediction of antigen-antibody interaction is essential for immune system research and treatment strategies. Experimental techniques for capturing the structural details of antigen-antibody complexes, however, can be extremely challenging and resource intensive. Computational modeling is a viable alternative, but current methods for computationally predicting such antigen-antibody docking are based on general protein-protein interaction studies, and only provide global scores of relative performances, without explicit structural rationales. We will demonstrate prototypical examples of how an explainable AI (XAI) based deep learning algorithm can be applied to determine antigen-antibody binding interfaces and interaction parameters. With viral envelope proteins as our study case, we have curated an extensive structural dataset of epitope-paratope complexes and associated binding affinities, and extensively teased out clustering features underpinning these interactions. We have deployed this dataset to enable our machine learning driven predictive models, applying an efficient graph neural network-based method to model docking interfaces in atomistic level. Detailed treatment of covalent bonds, non-bonded interactions, and secondary structural features in defining the interfacial graph extraction preserves the core invariants of antigen-antibody specific conformational information, leading to improved performance. Structural predictions are being further interpreted through XAI methods to understand the underlying sequence-structure-binding relations. This methodology not only gives us a handle over the structural bases of antigen-antibody recognition, but also enables predictive engineering of the interface towards improved therapeutic design.

197al Advancing Alchemical Free Energy Methods: Enhanced Flexibility, Parallelizability and Configurational Sampling

Wei-Tse Hsu, Valerio Piomponi, Pascal Merz, Giovanni Bussi, Michael Shirts

Over the past decade, alchemical free energy methods have been a popular choice for the computation of solvation free energies and binding free energies, given their ability to connect the end states of interest via nonphysical pathways free from unsurmountable free energy barriers. However, traditional methods like expanded ensemble (EXE) or Hamiltonian replica exchange (HREX) may not be efficient in accurate free energy calculations if the slowest degrees of freedom are largely orthogonal to the alchemical variable. In addition, there exist systems where traversing all alchemical intermediate states is challenging, even if alchemical biases (e.g., in EXE) or coordinate exchanges (e.g., in HREX) are applied. This issue is exacerbated when the state space is multidimensional, which can require seamless communications between hundreds of cores that current parallelization schemes do not fully support.

In this study, we present our recent efforts in addressing these two issues separately. First, we propose **alchemical metadynamics**, which allows explicitly biasing configurational collective variables in alchemical free energy calculations within the metadynamics framework. With test systems with varying complexity, we demonstrate that alchemical metadynamics captures metastable states not necessarily accessible in traditional free energy methods, thus capable of accurate free energy calculations. For the second issue, we introduce the method of synchronous **ensemble of expanded ensemble (EEXE)**, which periodically exchanges coordinates of different replicas of EXE simulations. We show that this method can compute free energies consistent with the estimates obtained in EXE and HREX, while offering much higher flexibility. Importantly, its planned successor, asynchronous EEXE, performs parallelized simulations asynchronously, which allows close communications between large numbers of adopted processors, and adaptive changes to the parameters of ensembles in response to data collected. These recent efforts can be easily integrated together in ways such as asynchronous ensemble of alchemical metadynamics, promising enhanced flexibility, parallelizability and configurational sampling for a wider ranges of systems.

(A) Alchemical metadynamics

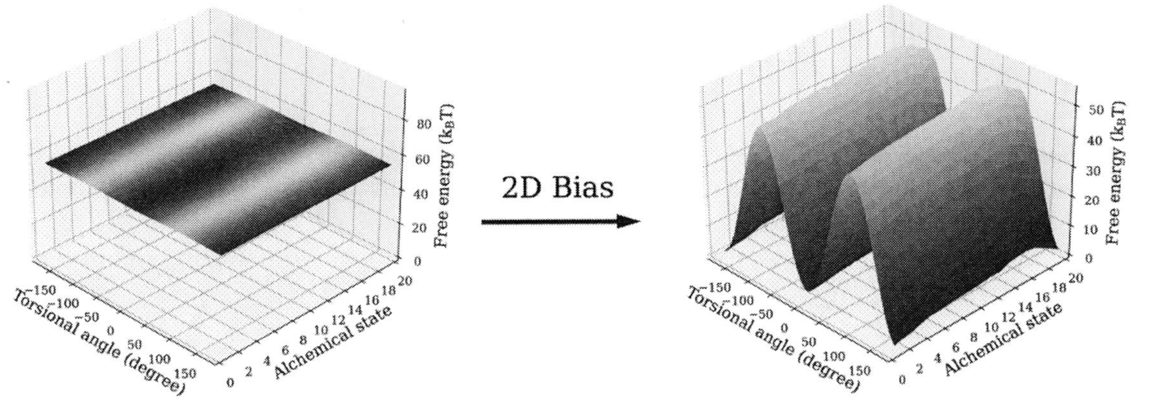

(B) Ensemble of exampled ensemble (EEXE)

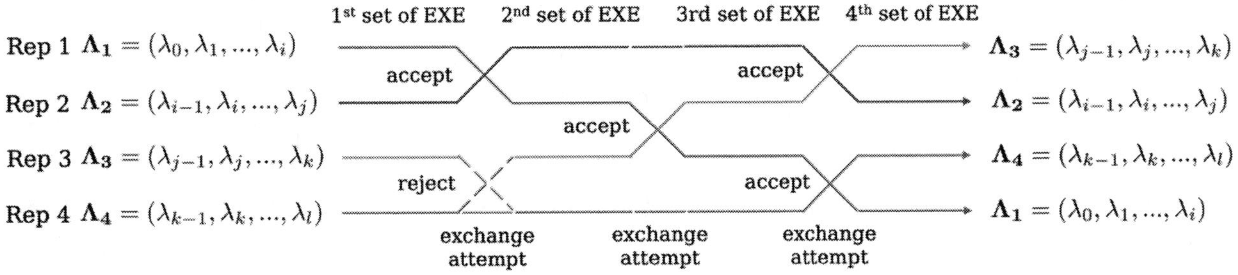

197am Determination of the Lipidated Structure of ApoE through Various *in silico* Methods

Emma Lietzke, Kimberley Bruce, Kayla Sprenger

Apolipoprotein E (ApoE), a major scaffold protein in lipoproteins, plays an important role in transporting lipids to microglial surface proteins to regulate brain lipid metabolism. ApoE has three isoforms (ApoE2/3/4), with ApoE4 considered to be the greatest known genetic risk factor for late onset Alzheimer's disease (AD). Despite much experimental evidence illustrating the impact of ApoE isoform on AD pathology, little *in silico* work has been established. This is largely due to the lack of available lipidated ApoE structures, as the only published crystal structure is of unlipidated ApoE3. Herein, we describe the use of two different computational approaches to obtain structures of the different ApoE isoforms in their lipidated forms, which we then rigorously compared to recently published experimental data. Specifically, we employed: (1) steered molecular dynamics (MD) simulations to bias the hinge regions of ApoE towards a conformation resembling ApoE in its lipidated form, and (2) homology modeling of ApoE against a published crystal structure of ApoA1 in a dimerized, lipid-bound, horseshoe-like conformation. A lipid nanodisc was then added to the structures using CHARMM-GUI, and MD simulations were performed to investigate the feasibility and stability of the structures in solvent. Overall, this work serves to elucidate the molecular-level changes that occur to ApoE as it becomes lipidated, paving the way for future *in silico* studies into the interactions of lipidated ApoE and many different proteins with prominent roles in neurodegenerative disease settings.

197an Adsorption of Adeno-Associated Virus Onto Ion-Exchange Chromatographic Media: A Coarse-Grained Molecular Dynamics Study

Tibo Duran, Shivangi Naik, Jake Kneski, Arani Chanda, Willow DiLuzio, Bodhisattwa Chaudhuri

Purpose:

Adeno-associated virus (AAV) packaging a single-stranded (ssDNA) genome has emerged as a promising vector and is the most common vehicle for *in vivo* gene transfer. They are well established in clinical trials for *in vivo* gene therapy with a good safety profile, low immunogenicity, and long-term stability. Currently, there is limited understanding regarding the mechanism involved in using anion-exchange chromatography to separate empty and full AAV capsids under different pH conditions during purification, as well as the capsids' propensity for adsorption onto the resin material and the role of loaded ssDNA in the adsorption process. Therefore, a coarse-grained molecular dynamics (CG-MD) simulations was conducted to probe the underlying mechanism of the adsorption and interactions between the capsid and strong anion exchange resin, such as polystyrene quaternary ammonium.

Methods:

The AAV Serotype 8 (AAV8) was chosen as the model viral capsid (PDB: 6v12) with loaded 2kb ssDNA built using the Avogadro software. The initial resin material was constructed and parameterized to the coarse-grained MARTINI force field and model. All the CG-MD simulations were conducted using the GROMACS software package and the trajectories were extracted and analyzed. The steepest descent algorithm was used in energy minimization followed by a 10 ns equilibration step in an isothermal–isochoric ensemble at ambient temperature. The simulation production runs were performed in an isothermal-isobaric ensemble using the Nosé–Hoover thermostat and the Parrinello–Rahman barostat at a pressure of 1 bar for each 10-fs time step. The simulations were carried out using periodic boundary conditions and performed computations in the High Performance Center Supercomputer Cluster at the University of Connecticut.

Results:

The adsorption behavior of capsid onto the anion-exchange chromatography resin surface at neutral and basic solution conditions was successfully determined using the CG-MD simulations. The analysis of the center-of-mass distance demonstrated that the full capsid bound to the surface more rapidly under basic conditions than other conditions. Conversely, empty capsids did not exhibit any adsorption under neutral conditions. Furthermore, the adsorption process was found to involve multiple asparagine and glutamine as the first amino acid residues to interact with the surface. Among the various conditions tested, the full capsid under higher temperature and basic conditions exhibited the highest level of instability. This instability was indicated by the higher root-mean-square deviation (RMSD) values and significant structural changes when binding to the resin, evidenced by the relative shape anisotropy. In addition, the full capsid under basic conditions had approximately 6% more anionic charge on its surface and exhibited much stronger binding interaction energy with the surface compared to the full capsid under neutral conditions.

Conclusions:

The CG-MD simulations enhanced our understanding of the adsorption mechanism between the AAV capsids and the anion-exchange chromatography. The developed computational model allowed us to evaluate the stability, electrostatic charges, and interaction energy of the empty and the ssDNA-loaded full capsids. Our simulations offered dynamic insights into the structure and energy of the capsids on a microsecond timescale, which revealed behavior that is unattainable through experimental investigations. The CG-MD simulations are demonstrated to be a powerful tool in the simulation of macromolecular complexes capable of guiding the interpretation, explanation, and direction of experiments.

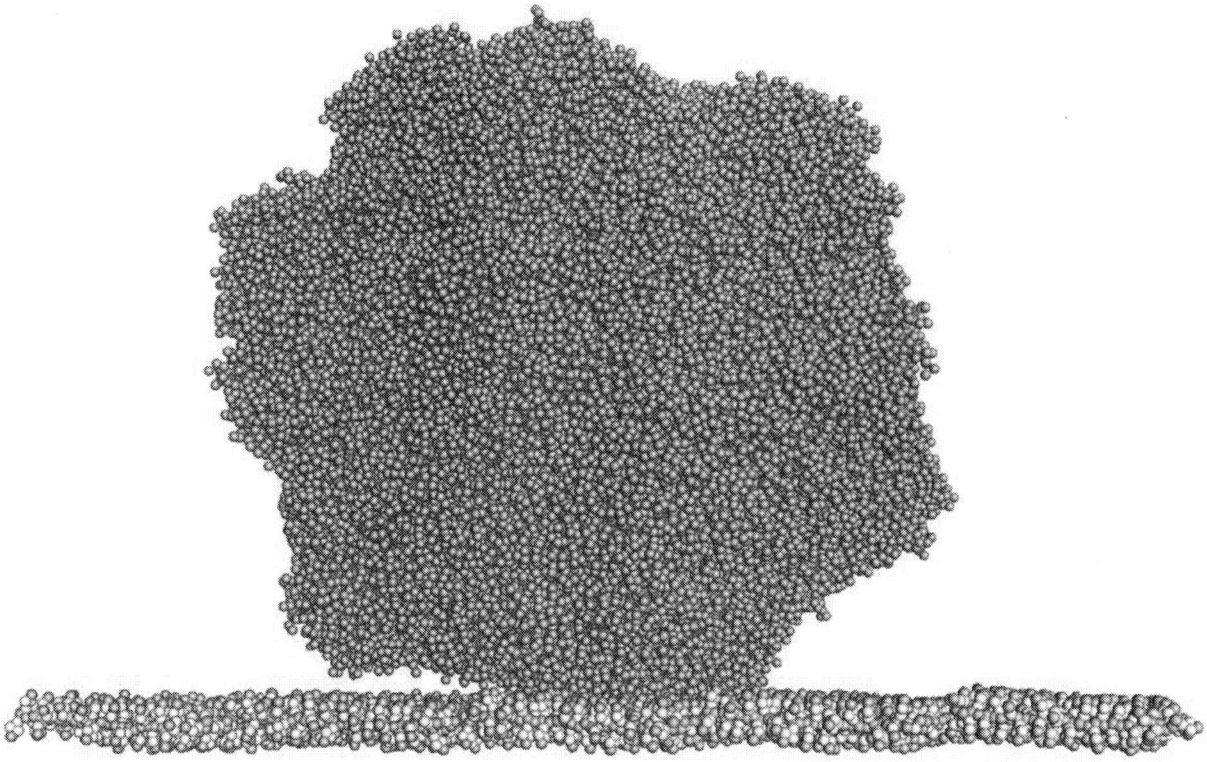

197ao Selective Agonist Design Targeting Cannabinoid Receptor 2

Soumajit Dutta, Diwakar Shukla

Cannabinoid receptors (CB1 and CB2) are important drug targets for inflammation, obesity, and other central nervous system disorders. However, drug design targeting these receptors has been a costly and time-consuming process with a high failure rate due to the off-target side effect. CB2 selective agonists can potentially treat pain and inflammation without the psychological side effects of CB1 agonism. In this project, with the combination of molecular simulation and data-driven modeling, we design selective molecules for CB2.

We hypothesize that the subtype selectivity of selective ligands can be explained by ligand binding to the conformationally distinct states between CB1 and CB2. ~700 μs of unbiased simulations are performed to study the activation mechanism of both receptors in the absence of ligands. Using neural network-based VAMPnets, we discretize the conformational ensembles of both receptors into metastable states. Structural and dynamic comparisons of these metastable states allow us to decipher a coarse-grained view of protein activation by revealing sequential conversion between these states. Specifically, we observe the difference in the binding pocket volume of different metastable states of CB1, whereas there are minimal changes observed in CB2. Docking analysis reveals that differential binding pocket volume leads to distinct binding poses and docking affinities of CB2 selective agonists in CB1. Only a few of the intermediate metastable states of CB1 shows high affinity towards CB2 selective agonists. On the other hand, all the CB2 metastable states show a similar affinity for CB2 selective agonists, explaining these ligands' overall higher affinity towards CB2. Furthermore, we performed a data-driven approach for designing CB2 selective ligands. The selectivity of the selective CB2 agonists is decided based on the 100-fold difference of the predicted binding affinity (pKi>2) between CB1 and CB2. The functionality of these molecules is determined based role of the molecules in activating the receptor (agonist/antagonist). We create a predictive data-driven workflow for selectivity using the existing database of cannabinoid assays for binding and function. The latent space obtained from the junction tree variational autoencoder (JTVAE) is used to featurize the molecules for predicted models. This VAE is pre-trained on all class A GPCR ligands (~160K molecules). Based on the stochasticity of the latent space, a data augmentation strategy is used to train prediction models. This strategy is shown to improve the accuracy of the prediction results significantly. The random forest algorithm is used with ten-fold cross-validations for regression and classification. The overall performance (R2 score) of our regression models is more than 0.96. With this network, the entire ZINC20 and ChEMBL databases will be screened. The screened molecules will be further validated by docking into CB1 and CB2 metastable states obtained from our simulations. Overall, with this computational study, we assist the CB2 drug discovery process by mechanistically explaining the subtype selectivity of CB2 selective ligands and designing a workflow for predicting selective molecules.

197ap First-Principles Study on the Role of Oxygen-Functional Groups in Heterogeneous Interface of MoS_2/Reduced Graphene Oxide for Sodium Ion Battery Anode

Wonmyung Choi, Byungchan Han, Sungjun Hong

Reduced graphene oxide (rGO) is used to improve the electrochemical performance of anode materials in sodium ion batteries by organizing rGO-based heterogeneous bilayers. However, it still lacks in-depth fundamental understanding of the heterointerface in these heterogeneous bilayers and how this affects their electrochemical performance. In this work, using different computational models (MoS_2/Graphene (Gr), MoS_2/Gr-O, MoS_2/Gr-OH, and MoS_2/Gr-COOH) we study Na affinity in the heterointerface and how the different oxygen-functional groups influence the sodium ion battery anode performance of the heterogeneous bilayers. Our density functional theory (DFT) results show that the major descriptor for the Na affinity in heterogeneous bilayers is unveiled as the non-covalent bonding nature, not charge distribution around Na ion. Furthermore, the Na affinity of each functional group at the MoS_2/rGO heterointerface is consistent with the trend of the activation barrier of Na. To infer the charge capacity, the phase stabilities are evaluated by adsorbing Na on a single layer of rGO, and the results show that Na-adsorbed rGO has more stable phases than graphene. The epoxy functional group enables maintains not only the charge capacity but also reduces the activation barrier of Na. These discoveries on the relationship between oxygen-functional of rGO in heterogeneous bilayer and their electrochemical performance will motivate the design of advanced anode materials for sodium ion batteries.

197aq Strategic Use of Molecular Simulations to Expand Predictive Capability of Machine Learning Models

Amey Thorat, Jindal Shah

Gas solubility in solvents can be estimated by calculating the excess chemical potential and determining the corresponding Henry's constant value. Rigorous computational methods implement the Bennett Acceptance Ratio approach to estimate the excess chemical potential. In this approach, the free energy change due to reversible dissolution of the gas is calculated by gradually introducing (or eliminating) the Coulombic and van der Waals' interactions between the solute and solvent in series of intermediate steps. This process is generally slow and computationally expensive and is inefficient at screening large chemical spaces in a high-throughput manner. Further, it can be used only for solvents with reliable force fields which severely limits its use. This calls for machine learning approaches to predict gas solubility. However, machine learning methods do not necessarily capture the underlying physics of the system and rely heavily on the quantity and quality of training data. In this work, we implement a hybrid approach to test if we can improve the predictive capability of machine learning models by strategically leveraging molecular simulations to expand the dataset. For example, extensive data is available for CO_2 solubility in linear alkanes, or linear alkyl substituted molecules in general, while data on branched alkanes is limited. In this work we investigate the feasibility of using rigorous molecular simulations on few simple branched chain molecules to generate data that can greatly expand the predictive capabilities of the machine learning model towards branched chain alkyl substituted molecules. We also identify the key features of such molecules whose data facilitates rapid improvement of the model.

197ar pH Adjustment Driven By Automation and Artificial Intelligence

Alexander Pomberger

Buffer solutions have tremendous importance in biological systems and in formulated products. Whilst the pH response upon acid/base addition to a mixture containing a single buffer can be described by the Henderson-Hasselbalch equation, modelling the pH response for multi-buffered poly-protic systems after acid/base addition, a common task in all chemical laboratories and many industrial plants, is a challenge. Combining predictive modelling and experimental pH adjustment, we present an active machine learning (ML)-driven closed-loop optimization strategy for automating small scale batch pH adjustment relevant for complex samples (e.g., formulated products in the chemical industry). Several ML models were compared on a generated dataset of binary-buffered poly-protic systems and it was found that Gaussian processes (GP) served as the best performing models. Moreover, the implementation of transfer learning into the optimization protocol proved to be a successful strategy in making the process even more efficient. Practical usability of the developed algorithm was demonstrated experimentally with a liquid handling robot where the pH of different buffered systems was adjusted, offering a versatile and efficient strategy for a pH adjustment processes. Finally, scalability of the approach was investigated to better understand the impact of different reactor dimensions and viscosity on pH adjustment.

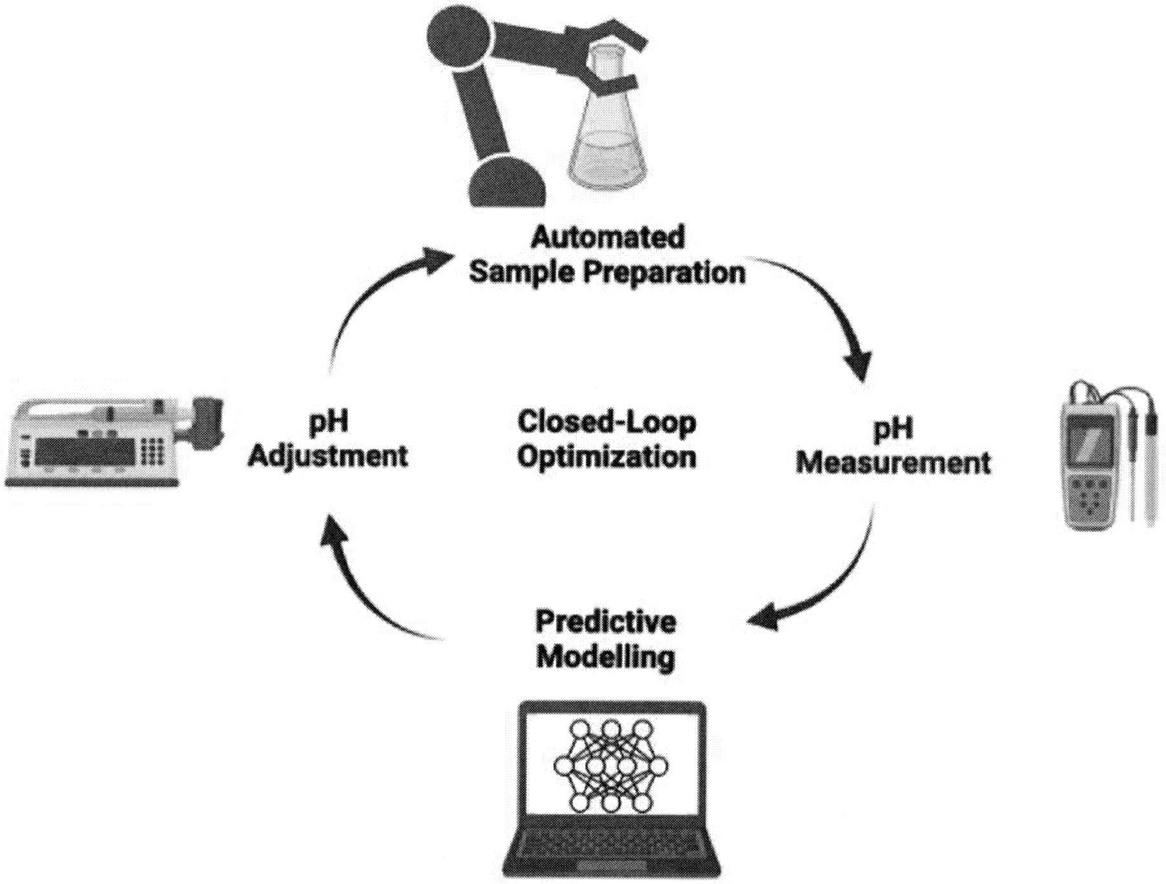

197at Assessing the Impact of Quantum Mechanical Descriptors on D-Mpnn Performance for Chemical Property Prediction

Shih-Cheng Li, Angiras Menon, Haoyang Wu, Kevin Spiekermann, William Green

Directed-Message Passing Neural Networks (D-MPNNs) are widely used for predicting chemical properties such as solvation thermodynamics, regioselectivity, and reaction barriers and energies However, the vast chemical space poses a challenge for such machine learning models to generalize well beyond the chemistry contained in the training set. One augmentation method that has received growing interest is the application of quantum-mechanical descriptors to D-MPNNs and other ML methods to improve their generalizability and performance. However, such quantum mechanical (QM) descriptors require expensive computational chemical calculations to compute, and it is not well-established which descriptors to use for different property tasks or under what data regimes these descriptors are most impactful. In this work, we perform a systematic study on the impact of QM descriptors on D-MPNN performance for predicting chemical properties across fourteen different datasets. The selected QM descriptors include atom-level properties such as partial charges, spin-densities, and shielding constants, bond-level properties such as bond-length, bond-order, and hybridization characteristics, and molecular-level properties such as HOMO-LUMO gap and multipole moment values. The target datasets include a variety of chemical properties ranging from electronic energy and solubility to toxicity and protein binding affinity, both classification and regression tasks, and a variety of dataset sizes spanning from several hundred to one-hundred thousand data points. We explore the effect of the level of theory, the choice of descriptors on specific property prediction tasks, how to best include descriptors in the D-MPNN framework, how down-sampling within large datasets changes the descriptors' impact, and differences between using directly computed QM descriptors and descriptors predicted by an additional D-MPNN as part of the workflow. We find that QM-descriptors can have a modest impact on the performance of D-MPNNs when training on small datasets or making predictions on difficult data splits where generalizability is key, but generation of additional data is often of greater benefit to model performance. Nonetheless, the improvement in small data regimes observed can be crucial when integrating ML models into *de novo* workflows for automatic molecular design and synthesis where experimental data is often limited. Generalizability is key in such applications as the targets are novel chemical species that are quite different to previously considered compounds. Additionally, even modest improvements in the accuracy of predictions are useful as false positives during synthesis are very expensive. In such applications, incorporating QM descriptors in conjunction with D-MPNNs can be a useful strategy for enhancing chemical property prediction and accelerating the autonomous molecular exploration workflow.

197au Multi-Fidelity Deep Learning for Data-Efficient Molecular Property Models from Experimental and Computational Data

Kevin P. Greenman, Temujin Orkhon, William Green, Rafael Gomez-Bombarelli

Chemistry and materials design frequently involve data of different fidelities to the true property to be optimized, for instance experimental measurements and computational approximations, or simulations at different levels of theory. The different fidelity levels typically involve a cost-accuracy trade-off, where high-fidelity results are relatively expensive or slow to acquire, while low-fidelity results are cheaper and faster but include more bias and/or noise in the data. Low-fidelity data is often useful for covering a broader portion of chemical space in high-throughput virtual screening campaigns in tiered screens (computational funnels). More recently, machine learning has been used to connect low- to high-fidelity data through sequential approaches such as transfer learning [1] and Δ-machine learning [2]. Multi-fidelity methods show some advantages to improve the generalizability of high-fidelity predictions. They are less expensive than Δ-machine learning since they also learn the low-fidelity function, avoiding extra cost at inference; and they leverage the two data fidelities at the same time (compared to transfer learning) [3]. However, many questions regarding when multi-fidelity methods should be expected to perform well remain unanswered, such as the relationship between dataset needs at both fidelities, and how the accuracy mismatch influences the results.

To address these questions, we present a comprehensive benchmark of multi-fidelity methods. We systematically add noise and bias to a synthetic dataset and split high- and low-fidelity data in ways that mimic realistic use cases. We also evaluate the multi-fidelity methods on several real-world datasets of optical properties (experiments and time-dependent density functional theory calculations), solubility (experiments and COSMO-RS calculations), and drug efficacy/potency (single-dose and dose-response measurements). We compare the multi-fidelity model performance to transfer learning and Δ-machine learning and provide recommendations for best practices in training models when multiple levels of fidelity are available. Finally, we demonstrate the application of uncertainty quantification and active learning in these models. The more thorough understanding of multi-fidelity methods we develop in this work will allow for more data-efficient molecular and materials design.

References:

[1]: Vermeire, Florence H., and William H. Green. 'Transfer learning for solvation free energies: From quantum chemistry to experiments.' Chemical Engineering Journal 418 (2021): 129307.

[2]: Ramakrishnan, Raghunathan, et al. 'Big data meets quantum chemistry approximations: the Δ-machine learning approach.' Journal of chemical theory and computation 11.5 (2015): 2087-2096.

[3]: Buterez, David, et al. 'Multi-fidelity machine learning models for improved high-throughput screening predictions.' ChemRxiv (2022).

197av Predicting Reaction Performance in Amide Bond Formation Using Machine Learning: The Role of High-Quality Data

Ye Seol Lee, Lingfeng Gui, Amparo Galindo, Claire S. Adjiman

Access to models that can predict the yield of organic reactions as a function of reaction conditions is of key importance as it can significantly accelerate the discovery and optimization of new chemical reactions. Recently, the rapid growth of artificial intelligence (AI) has led to the emergence of a variety of novel machine learning (ML) approaches, as efficient alternatives to time-consuming experimental screening, for reaction yield prediction [1,2] and reaction condition optimization [3,4]. Although the performance of such ML approaches and their potential roles in predictive chemistry have been demonstrated for specific reaction classes [1-4], their predictive accuracy is often limited by the lack of high-quality data; most of the publicly available data, such as USPTO and Reaxys, typically contain a limited subset of successful experiments at a small range of process conditions, and the concentrations of precursors, reaction times, temperatures and kinetics are only partially or not adequately reported [5]. As a result, the prediction models built upon these databases tend to be highly biased toward reactions with high yields.

In this work, we systematically investigate the impact of including unsuccessful reaction outcomes and reaction conditions in the development of ML-based reaction prediction models for amide bond formation. Generic data are extracted from the Reaxys database [6] to test the benchmark performance of the predictive models. In order to achieve accurate and unbiased results, a custom-made (CM) dataset for the amide bond formation is generated by sampling the generic reaction data such that the distribution of yield mimic those of the published high-throughput experimentation (HTE) data set [7]. Having curated the data, the ML-based models for predicting reaction yield [1] and optimizing reaction conditions [3] are trained. Finally, the comparative performance of such models is assessed to evaluate the importance of data quality and to investigate the applicability of CM data for building predictive reaction models.

References

[1] Schwaller, P., Vaucher, A.C., Laino, T. and Reymond, J.L., 2021. Prediction of chemical reaction yields using deep learning. Machine learning: science and technology, 2(1), p.015016.
[2] Ahneman, D.T., Estrada, J.G., Lin, S., Dreher, S.D. and Doyle, A.G., 2018. Predicting reaction performance in C–N cross-coupling using machine learning. Science, 360(6385), pp.186-190.
[3] Shields, B.J., Stevens, J., Li, J., Parasram, M., Damani, F., Alvarado, J.I.M., Janey, J.M., Adams, R.P. and Doyle, A.G., 2021. Bayesian reaction optimization as a tool for chemical synthesis. Nature, 590(7844), pp.89-96.
[4] Hickman, R.J., Aldeghi, M., Häse, F. and Aspuru-Guzik, A., 2022. Bayesian optimization with known experimental and design constraints for chemistry applications. Digital Discovery, 1(5), pp.732-744.
[5] Struble, T.J., Alvarez, J.C., Brown, S.P., Chytil, M., Cisar, J., DesJarlais, R.L., Engkvist, O., Frank, S.A., Greve, D.R., Griffin, D.J. and Hou, X., 2020. Current and future roles of artificial intelligence in medicinal chemistry synthesis. Journal of medicinal chemistry, 63(16), pp.8667-8682.
[6] Reaxys, Elsevier B.V.
[7] Avila, C., Cassani, C., Kogej, T., Mazuela, J., Sarda, S., Clayton, A.D., Kossenjans, M., Green, C.P. and Bourne, R.A., 2022. Automated stopped-flow library synthesis for rapid optimisation and machine learning directed experimentation. Chemical Science, 13(41), pp.12087-12099.

197aw ML-SAFT: A Machine Learning Framework for PCP-SAFT Parameter Prediction

Kobi Felton, Lukas Raßpe-Lange, Jan Rittig, Kai Leonhard, Alexander Mitsos, Julian Meyer-Kirschner, Carsten Knoesche, Alexei A. Lapkin

Fast and accurate prediction of fluid-phase thermodynamics is an important aspect of molecular discovery and process development. For example, solubility is an essential molecular descriptor in many drug discovery programs, and vapor liquid equilibrium is often a key consideration during the development of separation processes. Over the last fifty years, a variety of predictive thermodynamic methods have been developed ranging from group contribution methods[1, 2, 3] to quantum chemical simulations[4] to machine learning methods.[5, 6, 7, 8] However, there is still a need for methods that can extend to a wide range of compounds without significant tuning from the end user. Currently, group contribution methods require careful and often manual identification of predefined functional groups, while existing quantum mechanical (QM) methods often require significant expertise and computational cost. Machine learning methods have demonstrated promising results in the prediction of thermodynamic quantities such as activity coefficients, yet many lack the thermodynamic consistency of classical thermodynamic models.[9] Recently, it has been shown that using a machine learning model to predict the parameters of a Gibbs Excess model can overcome the challenges of including thermodynamic consistency inside neural networks.[10] We sought to apply the same concept of predicting parameters to an established Equation of State (EoS).

In this work, we develop ML-SAFT, a framework for predicting parameters of the PCP-SAFT EoS using machine learning.[11, 12] We were interested in the PCP-SAFT EoS because it can be used for a wide variety of thermodynamic prediction tasks including vapor liquid equilibrium,[11] solubility,[13] and surface tension,[14] yet each new molecule needs to be parametrized by regression to experimental data. To enable training of machine learning models, ML-SAFT includes the largest database of regressed PCP-SAFT parameters published in the literature (969 molecules) and a set of machine learning models trained on this dataset. We extract data from the Dortmund Databank[15] and develop a robust regression method to determine pure component PCP-SAFT parameters from experimental vapor pressure and liquid density data. Within ML-SAFT, we train random forests,[16] feed forward networks and message passing neural networks (MPNNs)[17] to predict the regressed PCP-SAFT parameters.

Our results show that random forests obtain the most accurate predictions of the regressed PCP-SAFT parameters. Furthermore, the best prediction of vapor pressure in terms of the average absolute deviation percentage (% AAD) on unseen molecules is obtained from the random forest. However, the best results on density predictions are obtained with parameters predicted by a MPNN. We attribute this difference to the increased representation capability of the MPNN for polar molecules, which we find to be important for density predictions. We also compare ML-SAFT models to two existing predictive PCP-SAFT models: SEPP[4] and group contribution PC-SAFT.[2] We find that ML-SAFT makes accurate predictions for a wider range of molecules than both methods while maintaining computational efficiency.

Overall, our work demonstrates that machine learning is a powerful tool for PCP-SAFT parameter prediction. We foresee that the results shown in this work can form a baseline for future work that explores multi-component mixture predictions using PCP-SAFT.

References

[1] A. Fredenslund, R. L. Jones, J. M. Prausnitz, Group-contribution estimation of activity coefficients in nonideal liquid mixtures, AiChE Journal 21 (6) (1975) 1086–1099. doi:10.1002/aic.690210607. URL https://doi.org/10.1002/aic.690210607

[2] E. Sauer, M. Stavrou, J. Gross, Comparison between a homo- and a heterosegmented group

contribution approach based on the perturbedchain polar statistical associating fluid theory equation of state, Industrial and Engineering Chemistry Research 53 (38) (2014) 14854–14864. doi:10.1021/ie502203w.
URL https://doi.org/10.1021/ie502203w
[3] D. Constantinescu, J. Gmehling, Further development of modified UNIFAC (dortmund): Revision and extension 6, Journal of Chemical and Engineering Data 61 (8) (2016) 2738–2748. doi:10.1021/acs.jced.6b00136.
URL https://doi.org/10.1021/acs.jced.6b00136
[4] S. Kaminski, K. Leonhard, SEPP: Segment-based equation of state parameter prediction, Journal of Chemical and Engineering Data 65 (12) (2020) 5830–5843. doi:10.1021/acs.jced.0c00733.
URL https://doi.org/10.1021/acs.jced.0c00733
[5] J. Habicht, C. Brandenbusch, G. Sadowski, Predicting PC-SAFT purecomponent parameters by machine learning using a molecular fingerprint as key input, Fluid Phase Equilibria 565 (2023) 113657. doi:10.1016/j.fluid.2022.113657.
URL https://doi.org/10.1016/j.fluid.2022.113657
[6] F. Jirasek, R. A. S. Alves, J. Damay, R. A. Vandermeulen, R. Bamler, M. Bortz, S. Mandt, M. Kloft, H. Hasse, Machine learning in thermodynamics: Prediction of activity coefficients by matrix completion, The Journal of Physical Chemistry Letters 11 (3) (2020) 981–985. doi:10.1021/acs.jpclett.9b03657.
URL https://doi.org/10.1021/acs.jpclett.9b03657
[7] E. I. S. Medina, S. Linke, M. Stoll, K. Sundmacher, Graph neural networks for the prediction of infinite dilution activity coefficients, Digital Discovery 1 (3) (2022) 216–225. doi:10.1039/d1dd00037c. URL https://doi.org/10.1039/d1dd00037c
[8] J. G. Rittig, K. Ben Hicham, A. M. Schweidtmann, M. Dahmen, A. Mitsos, Graph neural networks for temperature-dependent activity coefficient prediction of solutes in ionic liquids, Computers & Chemical Engineering 171 (2023) 108153. doi:10.1016/j. compchemeng.2023.108153. URL https://doi.org/10.1016/j.compchemeng.2023.108153
[9] K. C. Felton, H. Ben-Safar, A. Lapkin, DeepGamma: A deep learning model for activity coefficient prediction (2022).
[10] B. Winter, C. Winter, T. Esper, J. Schilling, A. Bardow, SPT-NRTL: A physics-guided machine learning model to predict thermodynamically consistent activity coefficients, Fluid Phase Equilibria 568 (2023) 113731. doi:10.1016/j.fluid.2023.113731.
URL https://doi.org/10.1016/j.fluid.2023.113731
[11] J. Gross, G. Sadowski, Perturbed-chain SAFT: an equation of state based on a perturbation theory for chain molecules, Industrial and Engineering Chemistry Research 40 (4) (2001) 1244–1260. doi:10.1021/ie0003887.
URL https://doi.org/10.1021/ie0003887
[12] J. Gross, J. Vrabec, An equation-of-state contribution for polar components: Dipolar molecules, AIChE Journal 52 (3) (2006) 1194–1204. doi:10.1002/aic.10683.
URL https://doi.org/10.1002/aic.10683
[13] M. Klajmon, Investigating various parametrization strategies for pharmaceuticals within the PC-SAFT equation of state, Journal of Chemical &amp Engineering Data 65 (12) (2020) 5753–5767. doi:10.1021/acs.jced.0c00707.
URL https://doi.org/10.1021/acs.jced.0c00707
[14] P. Rehner, J. Gross, Multiobjective optimization of PCP-SAFT parameters for water and alcohols using surface tension data, Journal of Chemical &amp Engineering Data 65 (12) (2020) 5698–5707. doi:10.1021/acs.jced.0c00684.
URL https://doi.org/10.1021/acs.jced.0c00684
[15] Dortmund databank (2022).
URL www.ddbst.com
[16] L. Breiman, Random forests, Machine Learning 45 (1) (2001) 5–32.
[17] J. Gilmer, S. S. Schoenholz, P. F. Riley, O. Vinyals, G. E. Dahl, Neural message passing for quantum chemistry, in: D. Precup, Y. W. Teh (Eds.), Proceedings of the 34th International Conference on Machine Learning, Vol. 70 of Proceedings of Machine Learning Research, PMLR, 2017, pp. 1263–1272.
URL https://proceedings.mlr.press/v70/gilmer17a.html

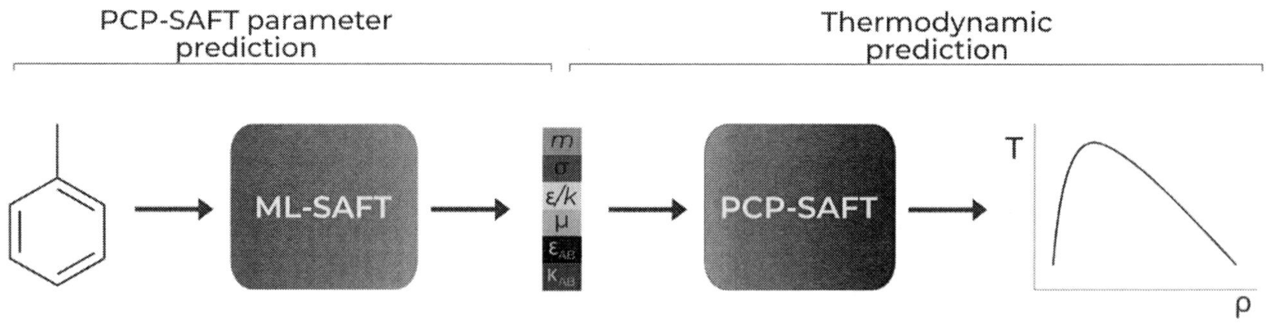

PCP-SAFT parameter prediction
Thermodynamic prediction
ML-SAFT
m
σ
ε/k
μ
ε_AB
κ_AB
PCP-SAFT
T
ρ

197ax A Universal Approach for Combining Computer-Aided Retrosynthesis and Retrobiosynthesis Tools

Xuan Liu, Hongxiang Li, Huimin Zhao

Enzymatic and chemical reactions have distinct synthetic spaces for designing synthetic pathways to target molecules due to the different characteristics between enzymes and chemical catalysts. A recent approach named ASKCOS (hybrid version) combines enzymatic and synthetic templates to obtain hybrid pathways, which demonstrates hybrid search can generate unique promising pathways. However, the use of template makes it only suitable for template-based synthesis planning tools. To address this limitation, we developed a Synthetic Field Score (SFScore) that can determine the appropriate synthetic field for single-step retrosynthesis of a target molecule. We hypothesized that if a compound has been reported in a chemical reaction corpus as a product, the compound is more favorable to be synthesized in the chemical synthetic field, and the same for the enzymatic reaction corpus. We trained a multilayer perceptron on a database of products extracted from an organic reaction database and an enzymatic reaction database. The model provides two scores for enzymatic and chemical synthetic fields, respectively. In the test dataset, the SFScore achieved average recall rates of 89.9%. The SFScore enables the search algorithm to discover more accessible intermediates and pathways within a time frame by avoiding unnecessary synthetic field searches, particularly for complex molecules that require long-step synthetic routes. The versatility of the SFScore allows template-free and template-based retrosynthetic tools to be combined, which leverages the strength of both approaches. In a hybrid search of 400 molecules, combining the template-free retrosynthesis tool named RetroTRAE with the template-based retrosynthesis tool named ASKCOS (enzymatic version) found hybrid pathways to 127 molecules, while entirely template-based ASKCO (hybrid version) found hybrid pathways to 202 molecules with the same terminating conditions. Although RetroTRAE's long search time prevented the pair from finding pathways for more molecules, the unique search space of the template-free tool allowed it to discover 30 molecules' pathways that ASKCO (hybrid version) could not find.

197az Dataset and Models for Predicting Critical Properties of Fluids

Sayandeep Biswas, Yunsie Chung, Haoyang Wu, William Green

Knowledge of critical properties, such as critical temperature, pressure, density, and acentric factor, is essential to calculate many different thermodynamic properties of chemical compounds. The values of critical properties are also used in models for mixtures, including for predicting temperature-dependent solvation energies and solubilities. Experiments to determine critical properties are expensive and time intensive, so it is desirable to glean as much information as possible from that hard-won experimental data. Therefore, we compiled/curated a data set of critical property values on 916 pure chemical compounds, and used that data to develop a machine learning (ML) model that can predict critical properties given as input the SMILES representation of a neutral chemical species. We explored directed message passing neural network (D-MPNN) and graph attention network as ML architecture choices. Additionally, we investigated featurization with additional atomic and molecular features, multi-task training, and pre-training using estimated data to optimize model performance. Our final model utilizes a D-MPNN layer to learn the molecular representation and is supplemented by Abraham parameters estimated using methods in the literature. A multi-task training scheme was used to train a single model to predict all the critical properties simultaneously with boiling point, melting point, enthalpy of vaporization, and enthalpy of fusion. The model was evaluated on both random and scaffold splits where it shows state-of-the-art accuracies. The extensive critical property data set containing 916 chemical compounds are made available in the public domain along with the model source code that can be used for further exploration.

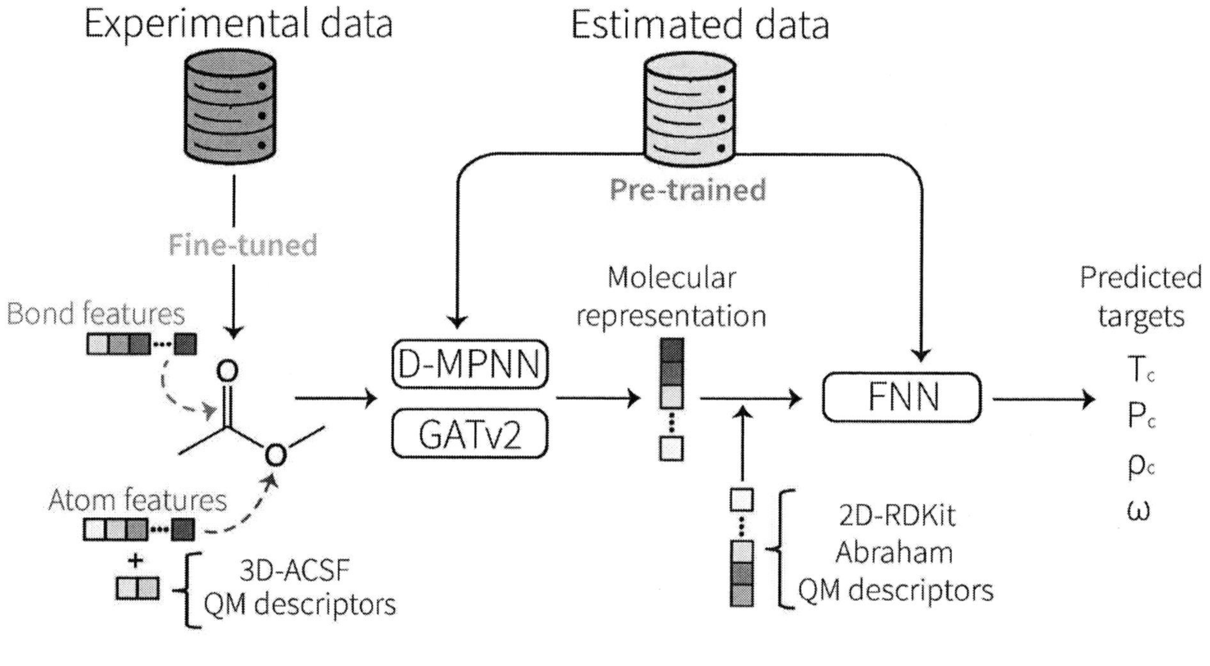

197bb Physics-Informed Bayesian Optimization Framework for Material Discovery

Oluwaseun Romiluyi, Maitreyee Sharma Priyadarshini, Kumar Miskin, Gigi Wang, Paulette Clancy

The lack of efficient discovery tools for advanced functional materials is a major bottleneck to enabling future-generation energy, health, and sustainability technologies. One main factor contributing to this inefficiency is the large combinatorial space of materials which is very sparsely observed. Searches of this large combinatorial space are often biased by expert knowledge and clustered close to material configurations that are known to perform well. Moreover, experimental characterization or first principles quantum mechanical calculations of all possible materials are extremely expensive leading to small data sets that are not suitable for a number of approaches, such as Deep Learning. As a result, there is a need for the development of computational algorithms that can efficiently search this large space for a given material application. Here, we introduce a method that combines a physics-informed belief model with Bayesian optimization. The material is characterized by physical and chemical properties of components of the material in a complex manner but *a priori* knowledge of the identity of the important properties is often lacking. The key contributing factor of our proposed framework is in the creation of a hypothesis space with all possible Gaussian process representations of the domain using these different elemental/molecular properties and the ability to select the hypothesis (belief model) that best represents our material design domain. The best hypothesis is then used to perform a search of the material space. Our method is unique since it picks out the physical descriptors that are most representative of the material domain making the search unbiased toward expert knowledge, which in many cases is unknown. The model also provides valuable chemical insight into the domain that can be used to develop new materials that were outside the domain that was initially searched.

197bc Expanding Bigsmiles for Automated Simulations and Machine Learning Representation of Polymeric Systems

Ludwig Schneider, Dylan Walsh, Bradley Olsen, Juan de Pablo

BigSMILES is a line notation used to describe complex polymeric systems. However, it does not describe a single system, but rather a family of possible realizations. To address this limitation, we present a compatible expansion of the BigSMILES line notation that adds information to make it generative.

With the expanded BigSMILES, the information from the line notation is sufficient to generate a full polymer system from scratch, including molecular weight distributions, percentages of monomers, reactivity and affinity of the monomers, as well as specifications such as solvents and mixtures. This enables the detailed atomistic generation of an ensemble of polymer molecules, providing a starting point for molecular dynamics simulations and the building of digital twins.

This expanded notation can also be used to generate initial conditions for high-throughput pipelines that analyze polymers using simulations and/or automated experimentation from a single line input prompt, combining completeness and human readability. Moreover, because the expanded BigSMILES describes exactly one ensemble of (random) polymer molecules, it is possible to determine the probability that a given molecule belongs to the described ensemble. This provides a starting point for training auto-encoders to represent these polymer ensembles for machine learning purposes, closing the loop between generating and quantifying molecules.

Overall, the expanded BigSMILES line notation provides a powerful tool for the design and analysis of polymeric systems, allowing for greater automation and efficiency in research and development.

197bd Expanding Chemical Synthesis Planning to Explore Chemo-Enzymatic Pathways Using Minimal Transitions

Vikas Upadhyay, Costas D. Maranas

Chemo-enzymatic pathway design involves chemical synthesis planning utilizing both enzymatic and non-enzymatic chemical reactions. Combining enzymatic and chemical steps can expand the chemical space accessible for synthesis beyond the capabilities of either set of reactions working independently. However, while molecules like tetrahydrofuran are straightforward to synthesize through chemical means, they can be challenging to produce using biological methods due to issues such as toxicity or a dearth of enzymes capable of catalyzing the necessary reactions. On the other hand, chemistry faces obstacles regarding regioselectivity and the ability to make alterations to complex and heavily functionalized molecules. Therefore, the chemo-enzymatic pathway design approach leveraging both chemical and biochemical reactions for synthesis planning enables the usage of enzymes to catalyze specific reactions and reduce the number of synthetic steps required to produce the target chemical. Recently, numerous studies have shown the advantages of leveraging enzymatic steps into industrial-scale chemical processes such as for the blood sugar regulator Sitagliptin (Merck) and anti-malarial drug Artemisinin (Amyris).

However, designing optimal chemo-enzymatic pathways is a complex task that involves encoding and understanding of chemical and biochemical reactions, as well as the knowledge of enzyme substrate activity and specificity. Furthermore, the task is complicated by the high-dimensional search space of potential reactions that combines both chemical/biochemical reactions and the need to optimize for factors such as maximizing yield, minimizing waste, and reducing the use of costly chemicals. Herein, we present a pathway design framework for exploring chemo-enzymatic pathways for chemical synthesis, connecting a starting chemical to the target molecule using both chemical and enzymatic reaction steps. Our approach minimizes the transitions between enzymatic and chemical reactions while designing pathways. The notion behind this concept is that executing these hybrid synthesis routes in experiments necessitates multiple stages of purification and separation of chemicals when there is a transition between enzymatic and chemical reactions. For instance, if a pathway involves a chemical reaction followed by an enzymatic reaction, it entails isolating and purifying the product of the enzymatic reaction to utilize it as the reactant for the subsequent chemical steps. Conversely, if the pathway involves a transition from a chemical to an enzymatic reaction, the purified product of the chemical reaction serves as the reactant for the subsequent enzymatic steps. Therefore, our goal is to reduce these transitions by identifying steps from both chemical and biochemical reactions to create pathways for chemical synthesis. To accomplish this, we use a mixed-integer linear programming algorithm to identify hybrid chemo-enzymatic pathways with the least possible number of such transitions. This approach enables us to investigate chemo-enzymatic routes that require minimal transitions, streamlining the planning of chemical synthesis using hybrid reactions and potentially leading to the development of more effective synthesis strategies. Additionally, our method provides the advantage of exploring an expanded chemical space for chemical synthesis, which could aid researchers in designing more efficient, sustainable, and eco-friendly synthetic routes.

197be Machine Learning-Enabled Modification of Polyamide Reverse Osmosis Membrane

Arash Tayyebi, Ali Alshami

Polymeric membrane design is challenging, and finding new polymer is time-consuming due to the complex interplay of various factors that influence the membrane's performance. Among the various membrane separation techniques, reverse osmosis (RO) is the most common architecture globally, offering excellent performance and simple operation for water desalination. Polyamide (PA)-based polymeric membranes are the current RO membranes in the market where their performance is typically evaluated by water permeability and salt rejection. There is a tradeoff relationship between permeability and water/salt selectivity and the goal of any modification is to push the boundaries of permeability/selectivity tradeoff. The present upper bound correlation defines a trade-off that empirically showcases the current state of the art in separation performance. Surface modifications (e.g., surface coating) is one of the most effective strategy to improve the performance of the PA-RO membranes.

Machine learning (ML) models as an effective and cost-saving tool can assist traditional trial-and-error processes to accelerate the discovery of polymers with anisotropic diffusional properties used in membrane fabrication. Though more number of studies have been performed on using ML to predict transport performance of polymeric membrane and accelerate the discovery of high-performance materials used in membranes, modifications of the PA layer of surface RO membrane has not been well explored with ML. To the best of our knowledge, there are still no ML model to modify polyamide layer with a reported ability to increase water permeability of RO membranes. To address this challenge, we explored the potential of ML to enhance water permeability of polyamide RO membrane and pushed the permeability/selectivity tradeoff across the boundary.

In this study, we trained an explainable screening ML model on a dataset of different polymers used as the active layer in RO membranes. By applying SHapley Additive exPlanations (SHAP) analysis to our model, we identified moieties with positive and negative contributions toward water permeability. We leveraged the results from the SHAP analyses to modify the PA layer of the commercial RO membrane by sharping the intensity of functional groups' peaks with positive effect and suppressing those with negative effect, resulting in a substantial increase in water permeability. The membranes were characterized using FTIR, XPS, SEM analyzer and tested for water permeance, and NaCl rejection using a dead-end stirred cell. Compared to the commercial membrane, the modified membrane improved the water permeance from 1.7 ml/30min to 3.4 ml/30min.

Our results demonstrate the potential of ML to replace the traditional trial-and-error to modify polyamide layer of PA-RO membranes for any particular separation application, paving the way for the development of more efficient and sustainable RO membranes for water treatment and purification applications. The proposed approach can be extended to other fields of materials science and engineering, where the identification of novel materials and/or modification of molecular structure with tailored properties is essential.

197bf Transition State Searches on Neural Network Potential Energy Surfaces

Jonah Marks, Joseph S. Gomes

Computational mechanistic studies of organic chemical reactions combine a model chemistry describing the potential energy surface on which chemical transformations take place together with algorithms for exploring these surfaces with the overall goal of producing reaction pathways, including the relative energies of all minima and potential energy barriers interconnecting them. Secondary goals include minimizing the computer time required to identify reaction pathways, and minimizing user intervention in the reaction pathway identification process. This is typically achieved through the use of local, surface-walking optimization techniques together with a nonlocal path-finding method such as the freezing string method.

Here, we describe the use of a neural network potential energy function together with nonlocal path-finding methods for guessing transition state structures connecting two minima. The neural network potential energy function is trained using a dataset of energies computed at the wB97X/6-31G* level of theory (ANI-1) for minimum energy structures of small organic molecules and off-equilibrium conformations obtained by normal mode sampling. We find that, though the calculation of energies along the reaction pathways is out of data distribution, the resulting model is able to qualitatively describe the potential surface along the coordinates of bond-making and breaking reactions.

We benchmark this methodology with a set of classic organic reactions and difficult cases. We show how transition state guess structures obtained with this method perform similarly to those obtained by electronic structure theory calculations in local surface-walking algorithms for saddle point searching, as measured by the number of electronic energy gradient calls required to successfully identify the transition state, with some exceptions. There is significant computational cost savings during the path-finding optimization by replacing the electronic energy gradients calls with gradients of the neural network potential energy function obtained by automatic differentiation.

197bg Automated Reaction Exploration of Solid Electrolyte Interphase

Hsuan-Hao Hsu, Brett Savoie

Solid electrolyte interphase (SEI) is a thin layer on the surface of an anode in a lithium-ion battery (LiB), serving as a protective layer that prevents the reactions between the electrolyte and anode. A well-formed SEI layer is essential for the stability and performance of LiBs. However, SEI degradation is a common problem in LiBs that causes reduced performance and safety issues. To understand SEI layer, SEI degradation has been extensively studied by experimental and computational methods. Due to the limitations of current experimental techniques, the exploration of reactions within SEI layer relies on computational methods, such as kinetic Mote-Carlo (KMC) method, and density functional theory (DFT). However, compared to experimental data, these reactions did not comprehensively cover all the products during SEI degradation. On the other hand, a more complicated reaction network by Persson was constructed by a machine learning model, which was trained by 5193 species and 86,001,275 reactions. However, this model still could not identify plausible species and reactions in SEI degradation. Here, an advanced reaction exploration program, Yet Another Reaction Program (YARP), is applied to explore the reaction network of LiBs. Different from using any filter based on chemistry knowledge in machine learning models, the graph-based elementary reaction step is utilized to explore all possible reactions in LiBs. Then, the transition states (TSs) are approximately located by the growing string method (GSM) with GFN2-xTB semi-empirical method (xTB). Finally, these TSs are refined by Berny optimization and intrinsic reaction coordinate (IRC) calculation to construct the reactions. With the idea of YARP, the reaction network with N-depth (N elementary reaction steps) can be constructed and the feasibility of all reactions in this network is guaranteed by quantum chemistry. The depth four exploration in ethylene carbonate (EC) shows that YARP is able to locate lithium ethylene mono-carbonate (LEMC), lithium ethylene di-carbonate (LEDC), and other byproducts that are consistent with experimental data.

197bh Study of Adsorption Behavior of Surfactant Micelles on Surfactant Covered Iron Surfaces Via Molecular Simulation

Abolfazl Faeli Qadikolae

Surfactant – metal interactions are important in applications like corrosion inhibition and synthesis of metallic nanoparticles, wherein surfactants are often added above their critical micelle concentration. Under these conditions, surfactant micelles exist in dynamic equilibrium with free surfactant molecules. Due to their faster diffusion compared to the micelles, free surfactant molecules rapidly adsorb onto metal-water interfaces. Therefore, adsorbing surfactant micelles invariably encounter metal-water inter-faces with some adsorbed surfactants. We have studied, via atomistic simulations, adsorption behavior of micelles of a quaternary ammonium-based surfactant at iron-water interfaces that are partially or fully covered with adsorbed surfactant molecules. We find that on a partially covered surface, a micelle prefers to adsorb on a bare metal patch. On a fully covered surface, the micelle adsorbs on top of already adsorbed molecules and attains a hemispherical morphology. The adsorption free energy decreases as the coverage of iron-water interface with surfactants increases.

197bi Evaluation of Polymer-Calcite Interfacial Strength through a Uniaxial Tensile Simulation Study

Keat Yung Hue, Myo Thant Maung Maung, Omar Matar, Paul F. Luckham, Erich Muller

***Please refers to the Abstract Figures image file for the figures caption label*

Introduction. In energy applications, when the carbonate rock is weak and poorly consolidated, the hydrocarbon extraction process produces undesired particles which will impact on hydrocarbon productivity and increase environmental waste. Solids production control is essential to mitigate the problem. The solids production risk can be reduced by the injection of formation-strengthening chemicals into the formation. By doing so, the formation compressive strength will be improved through the bonding of formation grains, thereby enhancing the intergranular forces. However, screening the formation strengthening chemicals can be experimentally tedious and time-consuming. In this research, molecular dynamics (MD) simulation was employed to screen different polymer candidates. This can be evaluated from the stress-strain response and the adhesion properties from a surrogate polymer-calcite uniaxial tensile simulation study.

Simulation Methods. The main component of carbonate formation, calcite, was represented with calcite crystal planes (1 0 4). Classical atomistic MD simulations and pcff+ forcefield were employed to model the interaction of the polymer with the calcite. Polyacrylamide-based polymers were selected as potential candidates, including pure polyacrylamide (PAM), hydrolysed polyacrylamide with 33% charge density (HPAM 33%) and sulfonated polyacrylamide with 33% charge density (SPAM 33%). As a control case, polyethylene (PE), a non-suitable candidate was also modelled as it was used extensively in the polymer-graphene surface system in literature [1] and served as comparison.

For each separate case, the polymer matrix was filled between the upper and lower solids surface layer, each consisted of 3 calcite layers. A series of equilibration steps were performed under microcanonical, isobaric-isothermal, and isothermal-constant pressure in z-direction ($NP_{zz}AT$) ensemble following a temperature quenching process from 500K to 300K at 0 atm. Before the tensile simulation process in the production run, a vacuum region was added on the calcite upper layer, where the upper and lower calcite layers were fixed rigid while the polymer was allowed to move freely as deformable region. In the tensile simulation, the upper layer was pulled upward in *z*-direction with a velocity of 0.0001Å/fs and the polymer deformed slowly for 800ps. The stress-strain response was measured from the normal force acting on the calcite lower layer normalized by the calcite surface area, and the polymer strength properties were analysed. In a separate simulation, the polymer matrix was frozen, and the stress-strain behaviour was measured with both the polymer and calcite upper layer detached from the lower calcite surface completely, at which the tensile stress measured was surrogate for polymer adhesion strength to the surface.

Results and Discussion. The stress-strain response behaviour is depicted in **Figure 1**, while the polymer deformation behaviour can be observed in the simulation snapshot in **Figure 2**. For the deformable polymer cases, in general, when the calcite upper layer is pulled upward, the polymer attached to the layer will be pulled together and deform slowly. The reluctance of the polymer to be deformed causes an increase in tensile stress, and when the displacement increases past a critical limit, a void is created in the bulk polymer region. At this stage, the polymer experiences stress failure: the polymer reaches peak tensile stress and decreases drastically due to the damage in the polymer matrix. Preliminary studies shows that the peak stress value recorded for the polymer-calcite system is the same as pure polymer system, which suggests the value is attributed to pure polymer tensile strength. This indicates the polymers can adsorb strongly onto the calcite surface, and the damage failure only occurs in the middle polymer region, but not at the polymer-calcite interface. Hence, the frozen polymer cases are modelled to force the detachment of the polymer from the interface, where the tensile stress recorded is considered as interfacial strength from literature [1].

Polyacrylamide-based polymers have amide functional groups that interact strongly with ionic calcite structure. The maximum tensile stress in **Figure 1** shows that polymer-calcite interfacial strength is significantly stronger than pure bulk polymer strength, validating the polymers can adsorb strongly to the calcite. In particular, the control case of PE has lower interfacial strength compared to other potential candidates. As shown in **Figure 2**, the PE has weaker adhesion and longer entangled chain away from the surface, while PAM system has stronger adhesion and complete adsorption to surface during the deformation process. Among the potential candidates, the HPAM33% shows better polymer strength due to the presence of carboxylate group in ionized form, which agrees with literature [2]. The results are further validated with the comparison to trend indication from the experimental results as shown in **Figure 3**, where the unconfined compressive strength (UCS) measured from the experimental study show similar trend to the polymers tensile strength in MD simulations.

Conclusion. Uniaxial tensile simulation studies allow the interpretation of the polymer deformation mechanisms and provide insights into the polymer-calcite interfacial strength analysis. It serves as a useful screening simulation tool to evaluate the suitable potential candidates for improved adsorption and formation strengthening performance.

References

1. Yuan, Z., et al., *A criterion for the normal properties of graphene/polymer interface.* Computational Materials Science, 2016. **120**: p. 13-20.
2. Thomas, M.M., J.A. Clouse, and J.M. Longo, *Adsorption of organic compounds on carbonate minerals. Model compounds and their influence on mineral wettability.* Chemical Geology, 1993. **109**: p. 201-213.

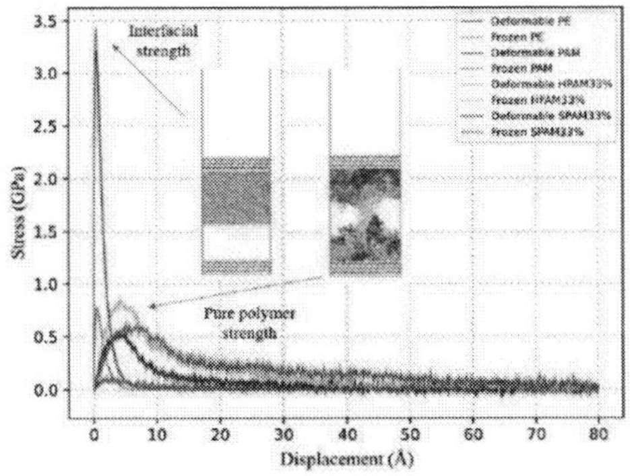

Figure 1. The stress-strain response behaviour of different polymer types during uniaxial tensile simulation for frozen polymer cases (dash line) and deformable polymer cases (solid line). Each polymer case was averaged from five different realizations.

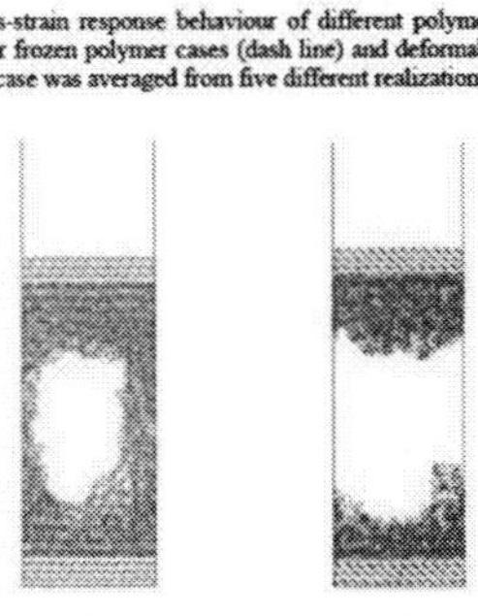

(a) (b)

Figure 2. Simulation snapshot of the polymer deformation behaviour in the uniaxial tensile simulation for deformable polymer cases. (a) PE. (b) PAM

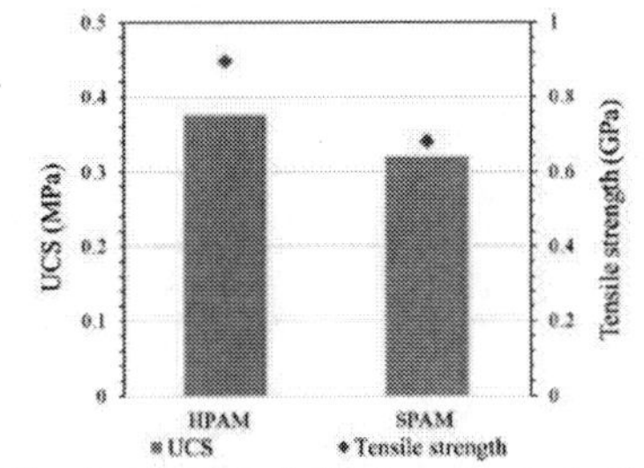

Figure 3. Comparison to trend indication from the experimental results. UCS refers to unconfined compressive strength measured from experimental study while tensile strength is measured from the bulk polymers strength from MD simulations.

197bj Relationship between Nanoscale Structure and Affinity for Organic-Modified Inorganic Solid/Organic Solvent Interface

Takamasa Saito, Ryo Takebayashi, Masaki Kubo, Takao Tsukada, Eita Shoji, Gota Kikugawa, Donatas Surblys

Nanomaterials including inorganic nanoparticles in liquids or polymers have been actively researched as highly-functional materials. For example, nanofluids are expected to be applied to coolant and conductive nano-ink, and polymer nanocomposites are expected to be applied to automobile parts, thermal interface materials, and optical thin films. To control the properties of these materials, it is important to control the dispersion state and spatial distribution of nanoparticles in host materials. Surface modification of nanoparticles with organic molecules is one of the effective technologies for controlling an affinity between nanoparticles and dispersed media, i.e., the dispersibility of nanoparticles. However, a design guideline for suitable surface modification is not well established because it is difficult to directly measure the affinity between surface-modified nanoparticles and dispersed media through experiments with high accuracy. In this study, molecular dynamics (MD) simulations were conducted for the interface between organic-modified inorganic solid and organic solvent to evaluate the interfacial structure and affinity.

All-atom MD simulations were performed for organic-modified inorganic solid substrate/organic solvent interfaces to simulate a part of a nanoparticle surface. The inorganic solid was α-Al_2O_3 (0001), the modifier was carboxylic acid with different chain lengths, and the solvent was *n*-alkane, also with different chain lengths. The surface coverage of modifier was 100% (4.8 molecules/nm^2) and 50%. The simulation systems were equilibrated at 300 K and 1 MPa. Work of adhesion was selected as a measure of interfacial affinity, and calculated by using the phantom-wall method, which is a kind of thermodynamic integration method. All simulations were conducted by using LAMMPS.

The effect of modifier chain length and solvent chain length on the work of adhesion was investigated. For the surface coverage of 100%, the work of adhesion remained almost unchanged because the solvent did not penetrate the modifier layer. For the surface coverage of 50%, the work of adhesion was maximum in the case of decanoic acid modifier. The longer the solvent chain length was, the larger the work of adhesion was. To evaluate the degree of solvent penetration to the modifier layer, the overlap parameter was calculated by using the density profile at the interface. The work of adhesion for various carboxylic acid-modified Al_2O_3/*n*-alkane systems was mostly correlated with the overlap parameter. Therefore, a high interfacial affinity can be achieved by appropriately selecting the modifier and the surface coverage to increase solvent penetration.

197bk Thermophysical Properties of H_2/Gas/Brine System Under Subsurface Storage Conditions: Molecular Simulations and Thermodynamic Modeling

Halla Kerkache, Hai Hoang, Pierre Cezac, Salaheddine Chabab, Guillaume Galliero

Deep saline aquifers seem to be a very promising solution for large-scale hydrogen storage (following a Power-to-Gas process) due to their capacity, their storage over a long period (providing sufficient storage capacity for seasonal energy demands) and their geographical availability [1, 2]. Given the limited knowledge of this type of storage, it is important to study the different phenomena that can impact the quality of the stored gas. In aquifers, the interaction between gas, brine and rock can lead to various physicochemical and biochemical phenomena that can impact the stability and mobility of hydrogen, including dissolution, diffusion and reactivity, etc. To study and develop such infrastructure, it is essential to have a comprehensive understanding and quantification of the thermophysical properties involved, such as gas/liquid solubility, diffusion, interfacial tension, and others [3].

For reasons of complexity, cost and hazardousness, experimental studies on the thermophysical properties of H_2/NaCl brine systems are limited. Few studies (2 experimental and 2 molecular simulation studies) were performed on the phase equilibria of H_2/NaCl brine systems [4-7], showing significant discrepancies, and only recently interfacial tension were evaluated [8, 9]. Regarding the transport properties, no experimental studies are available for the self-diffusivity of H_2/H_2O/NaCl mixtures, and only one recent molecular simulation study has been published [7]. Furthermore, almost no studies deal with more representative systems consisting of mixtures of gases (H_2, CH_4, CO_2, etc.) and ionic species in the formation water.

In this work, new equilibrium data for H_2/gas/brine systems are predicted using Monte Carlo simulation with different approaches [10]: Gibbs ensemble method (direct method) and single NPT simulation using Henry's law method (indirect method). The effect of density and excess properties on solubility prediction is studied using different force fields for hydrogen, gas, water and salt ions. In addition, the possible influence of clay interlayer on the solubilities is explored. Using molecular dynamics, new data on self-diffusivity and interfacial tension for H_2/NaCl brine systems are provided for a wide range of conditions relevant to underground hydrogen storage in porous media.

After validation against laboratory experiments, data are used to develop thermodynamic models with different approaches (gamma-phi and phi-phi approaches) and transport properties correlations. The developed models can be implemented in large-scale simulation tools, providing a more reliable representation of thermophysical property data and a better estimation of storage capacities and sealing integrity.

References

1. Raad, S.M.J., Y. Leonenko, and H. Hassanzadeh, *Hydrogen storage in saline aquifers: Opportunities and challenges.* Renewable and Sustainable Energy Reviews, 2022. **168**: p. 112846.
2. Muhammed, N.S., et al., *A review on underground hydrogen storage: Insight into geological sites, influencing factors and future outlook.* Energy Reports, 2022. **8**: p. 461-499.
3. Heinemann, N., et al., *Enabling large-scale hydrogen storage in porous media–the scientific challenges.* Energy & Environmental Science, 2021. **14**(2): p. 853-864.

4. Chabab, S., et al., *Measurements and predictive models of high-pressure H2 solubility in brine (H2O+ NaCl) for underground hydrogen storage application.* International Journal of Hydrogen Energy, 2020. **45**(56): p. 32206-32220.
5. Torín-Ollarves, G.A. and J.M. Trusler, *Solubility of hydrogen in sodium chloride brine at high pressures.* Fluid Phase Equilibria, 2021. **539**: p. 113025.
6. Lopez-Lazaro, C., et al., *Predicting the phase behavior of hydrogen in NaCl brines by molecular simulation for geological applications.* BSGF-Earth Sciences Bulletin, 2019. **190**(1): p. 7.
7. van Rooijen, W., et al., *Interfacial Tensions, Solubilities, and Transport Properties of the H2/H2O/NaCl System: A Molecular Simulation Study.* Journal of Chemical & Engineering Data, 2023.
8. Hosseini, M., et al., *H2− brine interfacial tension as a function of salinity, temperature, and pressure; implications for hydrogen geo-storage.* Journal of Petroleum Science and Engineering, 2022. **213**: p. 110441.
9. Higgs, S., et al., *In-situ hydrogen wettability characterisation for underground hydrogen storage.* International Journal of Hydrogen Energy, 2022. **47**(26): p. 13062-13075.
10. Kerkache, H., et al., *Solubility of H_2 in water and NaCl brine under subsurface storage conditions: measurements, molecular simulations and thermodynamic modeling. International Journal of Hydrogen Energy, (in Press, 2023).*

197bl Predicting Percolation in e-Waste Leaching Using a Coarse-Grained Molecular Flow Model

Zachary Diermyer, Yidong Xia, Jiaoyan Li, Ahmed Hamed, Jordan Klinger, Vicki Thompson

Recovery and recycling of electronic waste (e-waste) involves leaching to dissolve valuable metals as a key intermediate process. Preprocessing is required to mechanically size reduce and screen the raw e-waste to generate fine powders with improved metal concentration as feedstocks for leaching. The leaching efficiency largely depends on narrow particle size distribution (PSD), high sphericity and moderate surface wettability of the powders to form uniform pore networks and high-saturation percolation. However, the current e-waste leaching efficiency is not well understood or predicted, because of the poor powder qualities (wide PSD, large aspect ratio (AR) shapes, etc.); those result in heaps with preferential flow paths (short circuit) and low pore saturation, dissolving low metal content and requiring more re-circulations – a critical barrier for scaleup of the process. Improving leaching by using high-quality powders can be guided by computational simulations of percolation in realistic powder packing, which is relatively an unexplored area. Classical molecular dynamics (MD) simulations are not possible for the computing cost required for the relevant length and time scales. Continuum-based computational fluid dynamics (CFD) simulations are also not suitable, as they are not capable of resolving micron-scale pore-mediated wetting. This work uses a many-body dissipative particle dynamics (mDPD) model as a coarse-grained molecular flow model for multiphase simulations of percolation through pore networks formed by micron-sized e-waste powders. The model parameters are calibrated to exhibit the properties of realistic leaching fluids. Parametric simulation studies focusing on the influence of surface wettability and pore size and shape on the flux and saturation during percolation are performed. Criticality of the tested fluid properties and powder attributes as well as the processing parameters on the percolation performance will be reported and discussed.

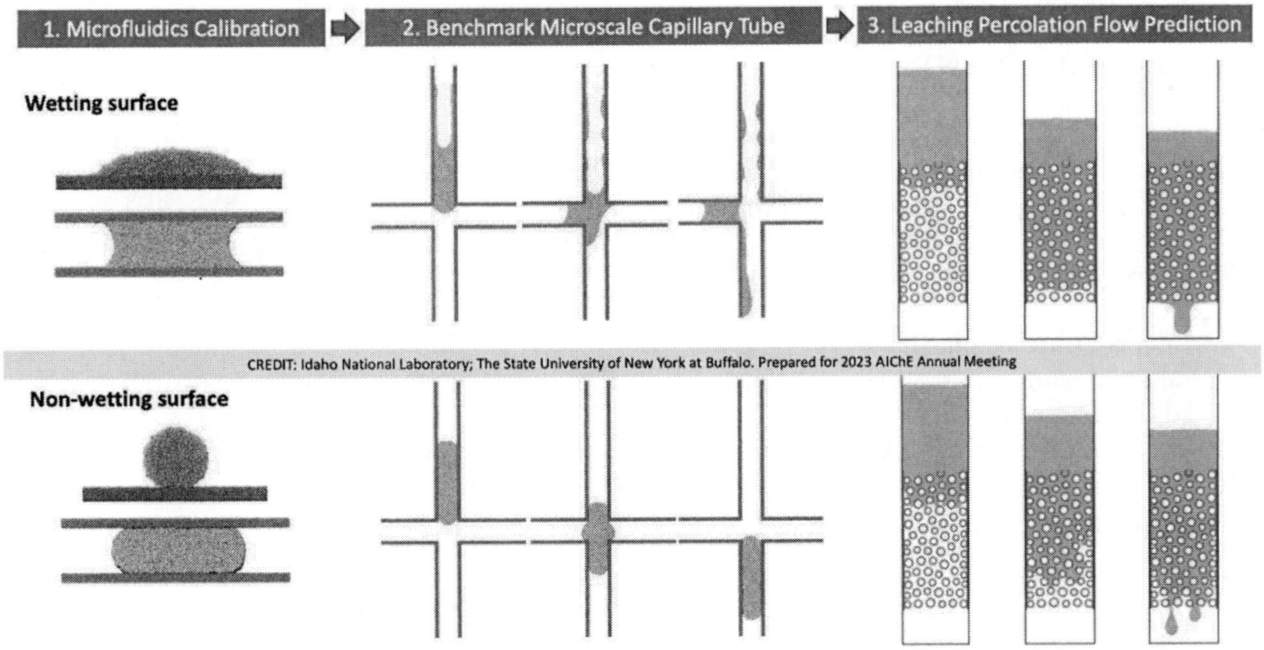

197bm Computational Investigation of Interfacial Transport in *Bijels*

Marco Tulio Portella, Xuan Duy Thao Nguyen, Dimitrios Papavassiliou

School of Chemical, Biological, and Materials Engineering

The University of Oklahoma, Norman, OK 73071

Bicontinuous Interfacially Jammed Emulsion Gels, also known as *Bijels*, find many applications in the industry including catalysis, separation processes and tissue engineering [1]. Recently, they have been considered as options for reactive separation media due to their large interfacial areas and ease of removal and insertion of reagents and products. In this work, polymer transport through *Bijel structure* is investigated through the use of Dissipative Particle Dynamics (DPD) computations, which is available through the Large-scale Atomic/Molecular Massively Parallel Simulator (LAMMPS) software package [2-5]. The two fluid phases are water and diethyl phthalate, stabilized with neutrally wetting nanoparticles. Surface parameters, such as interfacial tension and contact angle, are measured using the DPD coarse grained simulations in order to validate the computations and to guarantee *Bijel* stability [6]. Thereby, the interactions between the immiscible phases and the jammed nanoparticles can be evaluated. By the insertion of intra-bead interaction potentials, a multimeric molecule is modelled in DPD focusing on conformation identities such as the radius of gyration. The presentation will focus on the selection of the DPD model parameters that mostly affect the multimer conformation properties as a function of DPD beads. The discussion will include the calculation of the multimer diffusivity in each of the water and oil phases, and then the diffusivity across the oil-water interface. The mass transport coefficient as a function of the polymer-solvent interactions and the interfacial tension on the bijel system is obtained in order to model the migration of lipophilic compounds from the water to the oil phase.

ACKNOWLEDGEMENTS

The support of NSF under grant EFRI-2132141 is gratefully acknowledged as the use of computing facilities at the University of Oklahoma Supercomputing Center for Education and Research (OSCER) and at XSEDE (under allocation CTS-090025).

REFERENCES

1. Cha, S., Lim, H. G., Haase, M. F., Stebe, K. J., Jung, G. Y., Lee, D. "Bicontinuous Interfacially Jammed Emulsion Gels (bijels) as Media for Enabling Enzymatic Reactive Separation of a Highly Water Insoluble Substrate." *Scientific Reports* 2019, **9** (1), 6363.
2. Groot, R.D., Warren, P.B. "Dissipative particle dynamics: Bridging the gap between atomistic and mesoscopic simulation." *The Journal of Chemical Physics* 1997, **107**(11), 4423-4435.
3. Plimpton, S., "Fast Parallel Algorithms for Short-Range Molecular Dynamics." *Journal of Computational Physics* 1995, **117**(1), 1-19.
4. Nguyen, T.X.D., Vu, T.V., Razavi, S., Papavassiliou, D.V., "Coarse Grained Modeling of Multiphase Flows with Surfactants," *Polymers* 2022 **14**(3) Art 543 (19 pages).
5. Vu, T.V., Papavassiliou, D.V., "Synergistic effects of surfactants and heterogeneous nanoparticles at oil-water interface: Insights from computations," *Colloid & Interf. Sci.* 2019, **553,** 50-58.
6. Tavacoli, J. W., Thijssen, J. H. J. & Clegg, P. S. Bicontinuous emulsions stabilized by colloidal particles. RSC Soft Matter vols 2015-Janua (2015).

197bn A Computational and Experimental Characterization of the Ligand Environment of a Ni-Oxo Catalyst Supported in the Metal–Organic Framework Nu-1000

Stephen Vicchio, Zhihengyu Chen, Karena Chapman, Rachel Getman

Heterogeneous catalysts exhibit significant changes in composition due to the influence of operating conditions, and these compositional changes can have significant effects on catalytic performance. For bulk metal heterogeneous catalysts, relationships between composition and catalytic operating conditions are well documented, such as the relationship between coverage and structure with operating conditions on bulk metal surfaces. However, at present, the influence of operating conditions on the compositions of single-site heterogeneous catalysts remains largely unresolved due to the tunability and variability of the ligand environment of the active site. Single-site heterogeneous catalysts are highly attractive due to improved metal utilization and tunability; however, chemical intuition about the influence of operating conditions on compositions and structures of single-site heterogeneous catalysts hinders their design and rational optimization. To address this, we combined computational and experimental characterization of a Ni-oxo ($Ni4O_XH_Y$ clusters) catalyst under catalytic hydrogenation conditions. Pair distribution function (PDF) analysis is combined with *ab initio* thermodynamic modeling to investigate ligand environments present on a Ni-oxo cluster supported in the NU-1000 metal–organic framework. Experimental PDFs provide insights into both the Ni-O coordination numbers and the Ni–O, Ni···Ni, and Ni···Zr distances, which are key in understanding the molecular structure of the catalyst. These measurements are compared directly to simulated PDFs from Density Functional Theory (DFT) optimized calculations (a total of 854 different ligand environments). Our findings suggest significant OH and H_2O content, and that different Ni ions within the cluster and/or NU-1000 structure may comprise subtly different numbers of these ligands. The observation of significant H_2O content even under H_2 (g) conditions suggests that the support (NU-1000) supplies H_2O to the cluster. The combination of simulations and experiments provides new insights into the ligand environment for Ni-NU-1000 catalysts that will be useful for understanding the ligand environments of other single-site Ni catalysts as well.

197bo A Combined Experimental and Theoretical Study on Tuning Selectivity in Furfural Acetalization Reaction on Pd Nanostructures

Pallavi Dandekar, Govind Porwal, Tuhin Suvra Khan, M. Ali Haider, C.P. Vinod, Shelaka Gupta

Furfural acetalization reaction on well-defined Pd nanostructures showed structure sensitivity with respect to the particle shape. The acetalization reaction was performed over Pd cubes (~ 22 nm), octahedras (~ 26 nm) (Figure 1(a) and (b)) and spheres (~ 25 nm) at 303 K and 5 bar H_2 pressure for 3 hours in different solvents (ethanol, methanol, butanol and propanol). Pd cubes (exposing only the (100) facet) exhibited high furfural conversion (80%) and acetal selectivity (93%) compared to Pd octahedras (exposing only the (111) facets) (35% and 77%) and Pd Spheres (exposing both (111) and (100) facets) (41% and 75%) in ethanol as the solvent. The furfural acetalization reaction on Pd cubes was further studied in presence of methanol, propanol and butanol. The reactivity trend for different solvents followed the order: methanol (95%) > ethanol (80%) > propanol (62%) > butanol (56%). *ab-initio* Density Functional Theory (DFT) calculations further provided a mechanistic insight into the reactivity of two different Pd facets (100) and (111) towards furfural acetalization reaction at high coverage of solvents. DFT calculations showed that the binding energy of furfural on Pd (100) (E_b = -172.7 kJ/mol) (Figure 1 (c)) was higher than Pd (111) surface (E_b = -114.8 kJ/mol) (Figure 1(d)). On both the surfaces furfural acetal product was observed to be formed in three steps: alcohol dissociation into alkoxy intermediate followed by hydrogenation of carbonyl oxygen of furfural to form alkyl furan and nucleophilic addition of alkoxy group on the carbonyl carbon of alkyl furan to form acetal product (Figure 1(e)). The ethanol dissociation, furfural hydrogenation and acetalization step exhibited least barriers on Pd (100) (51.6, 27.7 and 77.2 kJ/mol) compared to Pd (111) surface (78.6, 35.8 and 92.2 kJ/mol) in presence of full coverage of ethanol. Calculated values of activation barriers for the three steps were significantly reduced on the Pd cubes in presence of methanol as the solvent (47.8, 24.2 and 64.6 kJ/mol) which was consistent with the experimental results.

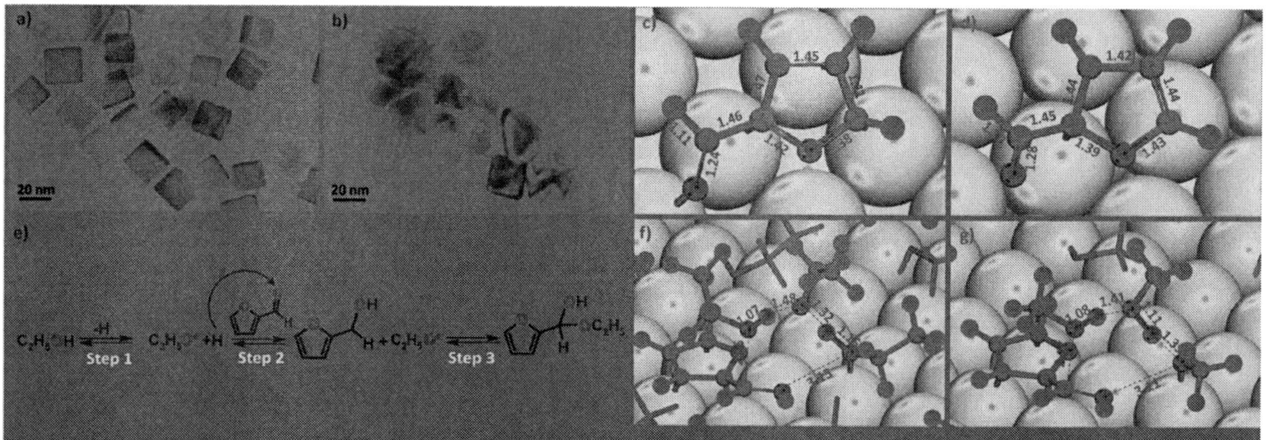

Figure 1: TEM image of (a) Pd cubes (~ 22 nm) (b) Pd octahedra (~ 26 nm), (c) binding modes of furfural on Pd (100) and (d) Pd (111) surface, (e) proposed reaction mechanism of furfural acetalization reaction, transition state structure of acetalization step on Pd (100) in presence of (f) ethanol and (g) methanol solvent.

197bp Computational and Experimental Studies of Selexol Solvent Physical Properties and Its Interactions with H_2o, H_2, and CO_2

Wei Shi, Robert Thompson, Jeffrey Culp, Surya Tiwari, Nicholas Siefert, Janice A. Steckel, David Hopkinson

Selexol is a physical solvent, which can be used to selectively remove acid gases, such as CO_2 and H_2S from a variety of natural and synthesis gas streams [1,2]. Selexol is very hydrophilic, which could absorb water from the gas stream and as a result to unfavorably decrease its performance in CO_2 capture [3]. In order to conduct reliable process modeling to analyze CO_2 capture in H_2 production using the Selexol solvent compared with hydrophobic solvents (such as diethyl sebacate [3]), thermophysical properties of Selexol and its interactions with other gas components, such as H_2O, CO_2, and H_2 are needed.

In this work, we studied various physical properties both from molecular simulations and experiments, such as gas absorption Henry's law constants and diffusivities for H_2O, CO_2, and H_2 in Selexol at 298-373 K, excess molar volume for solvent mixture, viscosities, and vapor pressures. Both simulations and the experimental data show that H_2O-Selexol solvent mixture exhibits the largest negative excess molar volume of 1.5-2.5 cm^3/mol at 298 K and a water mole fraction of 0.83. At this specific water mole fraction, large clusters formed between water and Selexol molecules were observed from simulations, which could maximize water-Selexol interaction and leads to the largest negative excess molar volume. In addition, At about this water concentration, both simulations and experimental data show that the water-Selexol mixture exhibit higher viscosity than at other water concentrations. Simulations show that CO_2/H_2 solubility selectivity in Selexol decreases from 88 ± 1 at 298 K to 39.2 ± 0.3 at 373 K, which are due to decreased CO_2 solubility however increase H_2 solubility at elevated temperatures. In the Henry's law constant region, both simulations and experimental data show that H_2O/CO_2 solubility selectivity is between 100-10000, indicating that Selexol is indeed very hydrophilic. At 298-373 K, simulations show that gas diffuses in the following relative order: H_2 (100) > CO_2 (10) > H_2O (1). Although Selexol and diethyl sebacate exhibit similar viscosities, both simulations and experimental data suggest that CO_2 diffusivities in diethyl sebacate is 4-5 times smaller than in Selexol. The vapor pressure for Selexol (a mixture of different chains of $(CH_2CH_2O)_n$) obtained from solvation free energy calculation was computed to be 0.02 ± 0.01 Pa at 298 K, comparable to the literature experimental value of 0.097 Pa. These findings suggest that Selexol is very non-volatile. In addition, at 298 K, the lightest component (CAS 112-49-2) is the predominant component in the gas phase; other heavier components have negligibly small concentrations in the gas phase. When the temperature is increased to 373 K, the saturation vapor pressure was estimated to be 22 ± 3 Pa, with a partial pressure of 19 Pa for CAS 112-49-2, and 3 Pa for the heavier CAS 143-24-8, respectively. Finally, calculations show that the solvation free energies for different components in Selexol correlate linearly very well with the number of repeat units for (CH_2CH_2O).

References:

1. Ashkanani, H. E.; Wang, R.; Shi, W.; Siefert, N. S.; Thompson, R. L.; Smith, K.; Steckel, J. A.; Gamwo, I. K.; Hopkinson, D.; Resnik, K.; Morsi, B. I. Levelized Cost of CO Captured Using

Five Physical Solvents in Pre-combustion Applications. *J. Greenh. Gas Control* **2020**, *10*, 103135.

2. Burr, B.; Lyddon, L. Which Physical Solvent is Best for Acid Gas Removal? *Process.* **2009**, *88*, 43–50.
3. Wei Shi, Surya P. Tiwari, Robert L. Thompson, Jeffrey T. Culp, Lei Hong, David P. Hopkinson, Kathryn Smith, Kevin Resnik, Janice A. Steckel, and Nicholas S. Siefert, Computational Screening of Physical Solvents for CO_2 Pre-combustion Capture, *Phys. Chem. B.* **2021**, *125*, 13467–13481.

197bq Computational Assessment of an Amine-Based Solvent Library for High-Salinity Brine Desalination

Gabriel Barbosa, C. Heath Turner

The growing population has led to a rise in demand for freshwater and other natural resources. TSSE, initially proposed in the 1960s for low-salinity brine desalination, has lately been found to be an effective method for desalinating high-salinity brine. Despite some attempts to study the TSSE performance of various amine-based solvents, the molecular properties of the best candidates for this solvent-based desalination method remain unknown. Using molecular simulation data and an unsupervised learning method, the molecular characteristics of 60 different amine-based solvents are analyzed. The results showed that the solvents are grouped into four clusters based on thermodynamic properties such as density, heat of vaporization, volumetric thermal expansion coefficient, solvent-solvent binding free energy, solvation free energy, and number of h-bonded sites. Additional brine-water interfacial simulations are conducted, and it is found that among the 59 amine-based solvents, secondary amines with a six-carbon branched or linear structure had the most promising TSSE performance.

197br Toward an End-to-End Computational Framework for Polymer Materials Design

Hanyu Gao, Yue Fang, Haifan Zhou

Polymer materials are extensively used in a wide variety of domestic and industrial applications, including energy storage, membranes, biomedical applications, etc[1,2]. Despite significant progress in new materials development over the past few decades, exploring the polymer space still holds strong potential due to the variability of monomer structure and polymerization methods, making the design space infinitely large. To meet the demand for better polymer materials with enhanced thermal, mechanical, optical, and electronic properties, as well as ease of synthesis and production, computational modeling has been used to simulate polymerization and polymer molecules, and can be used to support experimental efforts by providing guidance for more rational experiment design[3–5].

However, traditionally, polymerization modeling and polymer molecular simulation have been treated as separate subjects, resulting in suboptimal or infeasible design. Thus, an integrated computational framework for closed-loop polymer design is highly desirable, yet challenging to achieve due to the gaps in information and time scale of different simulation methods.

In this work, we present our efforts towards an integrated framework for designing polymerization recipes and targeting polymer material properties. We improved the kinetic Monte Carlo algorithm and combined it with an active learning algorithm to design the polymer chain sequence distribution efficiently. Coarse-grained molecular dynamic simulation was then used to connect the polymer chain information with the thermal and mechanical properties of the polymer material. Additionally, machine learning algorithms were employed to approximate molecular dynamic simulations and better unite the time scales of polymerization modeling and polymer molecular simulations. Our work is a first step towards an end-to-end computational framework for polymer materials design, with the potential to extend the scope of materials and connect with experiments.

Reference

(1) Ali, U.; Karim, K. J. B. A.; Buang, N. A. A Review of the Properties and Applications of Poly (Methyl Methacrylate) (PMMA). *Polym. Rev.* **2015**, *55* (4), 678–705. https://doi.org/10.1080/15583724.2015.1031377.

(2) Frontiers in Polymer Science and Engineering. *A Natl. Sci. Found. Spons. Work.* **2016**.

(3) de Pablo, J. J.; Jackson, N. E.; Webb, M. A.; Chen, L.-Q.; Moore, J. E.; Morgan, D.; Jacobs, R.; Pollock, T.; Schlom, D. G.; Toberer, E. S. New Frontiers for the Materials Genome Initiative. *npj Comput. Mater.* **2019**, *5* (1), 41.

(4) Gao, H.; Waechter, A.; Konstantinov, I. A.; Arturo, S. G.; Broadbelt, L. J. Application and Comparison of Derivative-Free Optimization Algorithms to Control and Optimize Free Radical Polymerization Simulated Using the Kinetic Monte Carlo Method. *Comput. Chem. Eng.* **2018**, *108*, 268–275. https://doi.org/10.1016/j.compchemeng.2017.09.015.

(5) Gao, H.; Oakley, L. H.; Konstantinov, I. A.; Arturo, S. G.; Broadbelt, L. J. Acceleration of Kinetic Monte Carlo Method for the Simulation of Free Radical Copolymerization through Scaling. *Ind. Eng. Chem. Res.* **2015**, *54* (48), 11975–11985. https://doi.org/10.1021/acs.iecr.5b03198.

197bs Prediction of Isomorphs By Using Configurational Temperature

Kelly Badilla, Andreas Bommarius, Marcus T Cicerone, Lorenzo Costigliola, Thomas B. Schrøder, Jeppe Dyre

When performing molecular dynamic simulations, lowering the time and energy spent on calculations can be very valuable. One way of reducing time spent on simulations is by eliminating the need of simulating systems at different conditions, which can be accomplished by identification of isomorphs[1]. Isomorphs refer to a set of equivalent state points in which structure and dynamics are invariant. They effectively reduce a two-dimensional thermodynamic phase diagram to one dimension; this allows for the mapping of properties of a system at one point to another point without need of running a simulation at the other point. Currently, there are a number of methods used for isomorph identification, but the focus of the presented work is on the potential of using configurational temperature, T_{conf}[2], for isomorph identification.

The most fool-proof method of isomorph identification is the gamma (γ) method, in which the main property of an isomorph is employed – constant excess entropy along the isomorph. Although reliable, this method is expensive as it requires a simulation from which γ can be calculated, γ being the correlation coefficient between the potential energy function and the instantaneous virial[1]. There are other methods for isomorph identification, all of which have varying degrees of speed compared to the γ method, but the focus of this work is on exploring the viability of using configurational temperature, T_{conf}, for isomorph identification, which requires only one equilibrated configuration from one simulation. To this end, we have seen success with single-particle Lennard-Jones systems, but the presented work expands to include one, two, and three component systems.

1. Chem. Phys. 131, 234504 (2009)
2. Phys. 103:10, 1361-1373 (2005)

197bu A Molecular Dynamics Study of Growth and Dissociation of Mixed Methane + Carbon Dioxide Gas Hydrates

Soumya Chatterjee, Bhavesh Moorjani, Jhumpa Adhikari

Gas hydrates are crystalline structures that are made of water (host) cages encapsulating the gas (guest) molecules. Methane hydrates are mostly found in the deep sea and permafrost regions; and are a source of fossil fuel where the estimated amount of carbon content trapped in hydrate reservoirs globally is twice that present in conventional fossil fuels[1]. Methane recovery from hydrate reservoirs using CO_2 injection has gained attention in the past few decades owing to its potential to suffice energy requirements and sequester carbon dioxide[1]. However, due to differences in hydrate forming conditions between methane and carbon dioxide hydrates full recovery of methane from the reservoirs cannot be achieved, as result residual quantities of mixed (CO_2 + CH_4) hydrates, CH_4 hydrates, CO_2 hydrates, CH_4 and CO_2 gases may remain in the hydrate reservoir. Although methane hydrates and carbon dioxide hydrates have been extensively studied in the literature, not many studies are available demonstrating the behavior of mixed gas hydrates[2], [3]. Therefore, in our work, we are studying the effect of various parameters, such as salinity, CO_2 concentration in the aqueous liquid phase, and temperature which shall affect the stability of mixed gas hydrates in the reservoir. To achieve this, we have performed molecular dynamics simulations (by employing GROMACS open source software package) of the mixed hydrate in contact with liquid water (H-L_w phase) in the NpT ensemble at a fixed pressure of 30.5 bar and temperatures of 250 and 273 K. We have observed the effect of addition of salts in the aqueous phase, supersaturation of CO_2 in the aqueous phase, and combined effect of both, on mixed hydrate behavior at both 250 and 273 K. Simulations have been performed using all-atom forcefields for water (TIP4P/2005), carbon dioxide (TraPPE Flexible), methane (TraPPE-EH Flexible) and NaCl to capture a reliable and effective description of the interactions present in the system. At low temperatures (250 K) in the presence of supersaturated CO_2 in the aqueous phase, growth of CO_2 hydrate has been observed. Order parameter analysis has been used to estimate the variation in the thickness of the hydrate with time and eventually the velocity of the hydrate front during growth has been calculated. The interface width at both interfaces has been observed to be invariant with time. However, at the same low-temperature condition, when salt was present along with supersaturated CO_2 in the aqueous phase, the growth of the hydrate is limited up to a few nanoseconds, and eventually, gas nanobubbles were formed. This confirms the role of salts in inhibiting the growth of the mixed hydrate. Hydrogen bonding and local order parameter analysis suggest that salts disrupt the hydrogen bonding between water molecules required to form the hydrate cages, thereby inhibiting the growth of the hydrate. Two-dimensional mass density profiles are used to understand the mechanism of nanobubble growth by measuring its change in radius with time. At low-temperature conditions, when neither CO_2 nor salts were present in the aqueous phase, the formation of ice was detected. However, when only salts and no CO_2 were added to the aqueous phase at the same temperature condition, no significant drop in the fraction of solid water was observed. This result is interesting by itself as it suggests a possible difference in the effect of salts in the presence and absence of an aqueous solution supersaturated with CO_2 at low temperatures. At the high-temperature condition (273 K), dissociation of the mixed hydrate was observed. However, when the liquid water phase was supersaturated with CO_2 at the same condition, a cylindrical nanobubble was observed. This is due to decreased solubility of CO_2 in the aqueous phase at high temperatures. Here, we attempt to explain the role of a mixture of guest molecules in the hydrates using interaction energies, the possible difference in the mechanism of growth and dissociation by studying interfaces, and the role of salt concentration in growth and dissociation by looking at hydration shells around ions and orientation of water molecules in the hydration shells. The work presented here will help to gain molecular insights for efficient methane recovery from and carbon dioxide sequestration in the hydrate reservoir by carbon dioxide injection.

References

[1] W. N. Wei, B. Li, Q. Gan, and Y. Le Li, "Research progress of natural gas hydrate exploitation with CO2 replacement: A review," *Fuel*, vol. 312, p. 122873, Mar. 2022, doi: 10.1016/J.FUEL.2021.122873.

[2] S. Sarupria and P. G. Debenedetti, "Homogeneous Nucleation of Methane Hydrate in Microsecond Molecular Dynamics Simulations," *J Phys Chem Lett*, vol. 3, no. 20, pp. 2942–2947, Oct. 2012, doi: 10.1021/jz3012113.

[3] S. Sarupria and P. G. Debenedetti, "Molecular dynamics study of carbon dioxide hydrate dissociation," *Journal of Physical Chemistry A*, vol. 115, no. 23, pp. 6102–6111, Jun. 2011, doi: 10.1021/jp110868t.

197bv Molecular Insights from Simulations on MRI Contrast Agents in Different Chemical Environments

Thiago Jose Pinheiro Dos Santos, Dilip Asthagiri, Philip Singer, Walter Chapman

In Magnetic Resonance Imaging (MRI), the nuclear magnetic resonance (NMR) relaxation of protons in water is used to probe the state of bodily tissues. The contrast in these images is enhanced by Gadolinium-based contrast agents (GBCAs) that act by reducing the water proton relaxation time. However, despite the long history of NMR, the molecular-scale processes in NMR and MRI remain poorly understood; the modeling and interpretation of the physics of relaxation still rely on severe assumptions. Further complicating matters, different tissues present different relaxation times due to variation in the physical and chemical environment, including nano-confinement, presence of osmolytes, electrolytes, and dissolved oxygen; the effect of such complexity is also not understood. In this work, we employ quantum and molecular simulations to investigate the effect of osmolytes on the NMR relaxation of gadolinium(III)-aqua complex and Gadavist®. Our simulation results are corroborated and validated by NMR relaxation dispersion measurements. We investigated the roles of urea and TMAO; urea is a well-known denaturant of protein structure, while TMAO has a stabilizing effect on protein structures. Our investigations show that such osmolytes can influence the NMR relaxation of gadolinium(III) complexes. In this presentation we will discuss these results. Further, we will discuss how the NMR relaxation signal can be decomposed into contributions from dynamical "molecular modes," a technique that can prove powerful in building a library of signals to interpret NMR and MRI signals in otherwise hard to analyze environments.

197bw Force Fields and Initial Conformations Optimization of Amorphous Polymers in Multi-Scale (atomistic and coarse-grained) Simulations.

Sahar Zenoozi, Peter J. Ludovice, Clifford Henderson

Polymers are one of the most diverse classes of materials thanks to the nearly infinite number of combinations of structures possible due to a variety of characteristics of polymer synthesis including the ability to utilize numerous different polymerization approaches available today, the wide array of monomer chemistries and architectures that are easily synthesized, the numerous polymer architectures beyond linear homopolymers that are available through various techniques such as the use of branching chemistries, and the relative ease of enchaining various combinations of different monomers along a polymer chain. Although this flexibility to exquisitely control and tailor polymer structure makes this class of materials well-suited to solve a variety of engineering challenges across a variety of application areas, the size of the polymer design space also presents a significant challenge when tasked with determining the best polymer structure to produce that can satisfy the property and processing requirements for a particular problem. With the dramatic gains in computational capability in recent years, molecular modeling has finally transitioned from a tool that can be used for predictive design and optimization of polymer structure as opposed to most past use cases where it was best suited to explore either general polymer behavior or to semi-quantitatively match and explain experimental data for real polymer systems in hindsight. Thus, the ability to develop of reliable predictive computational models for the behavior and properties of real polymer systems is of great importance for the current and future design and optimization of new materials. In this work, a methodology is presented which can be used to develop improved atomistic and coarse-grained force fields for molecular dynamics simulation of real polymers. Furthermore, an efficient method for producing initial conformational states in dense linear polymer simulations is also presented.

The first challenge when creating polymer molecular dynamics simulations that are expected to quantitatively match, and better yet predict, polymer properties is the selection of a force field to represent the interactions of the atoms and/or coarse-grained structural elements of the polymer. There is in general always a need for the ability to conduct accurate atomistic MD simulation of polymers. In some cases, such atomistic simulations can be useful themselves and in other cases they can be useful for parameterizing and training coarse-grained models that can allow for larger time and spatial scale simulations. In this work, an improved force-field is developed for the simulation of linear hydrocarbon polymers based on initial use of the OPLS force-field with various refinement and optimization approaches. The end result, as will be demonstrated from the results presented for several exemplar linear homopolymers including polystyrene (PS), polypropylene (PP), and polyisoprene (PI), is a force-field which can reproduce the physiochemical properties of such linear hydrocarbon polymers with significantly improved accuracy. Once such improved atomistic force-fields and simulation were available, these capabilities were then utilized to develop improved coarse-grained models for such linear hydrocarbon polymer systems. In particular, models that allow for higher levels of coarse graining while retaining the ability to predict important polymer structural and physical properties were explored and developed. The general methodology used for this coarse-grained model development will be presented and the specific coarse-grained models developed for PS, PP, and PI will be presented along with a discussion of the ability of such coarse-grained models to accurately reproduce the behavior and properties of these polymer systems.

An often underappreciated challenge in such polymer modeling problems is the ability to generate realistic polymer conformations for polymers at their experimental density. This paper will also discuss the importance of such polymer conformations with respect to the general ability to produce accurate molecular dynamics models for linear polymer systems that can reproduce the properties of well characterized polymer systems (i.e. PS, PP, and PI). Improved methodologies for generating initial conformations for use in atomistically detailed simulations of dense polymer systems will be discussed and benchmarked in detail.

The end result of this work is a set of methodologies and both atomistic and coarse-grained polymer simulation models that can reproduce a wide variety of the physiochemical properties of linear hydrocarbon polymers. The future potential to extend this work to both other known and as of yet to be synthesized polymer systems will be discussed.

197bx Entropic Stabilization Plays a Key Role in the Non-Uniform Distribution of Oxygen Ions and Vacancy Defects in Gadolinium-Doped Ceria

Bharathi Bandi, Abhijit Chatterjee, Methary Jaipal

Gadolinium-doped ceria (GDC) is an important fast oxygen ion conductor. O^{2-}-ion hopping in different cation environments of GDC is probed using molecular dynamics (MD). We find that at equilibrium, the O^{2-}-ions and vacancy defects arrange themselves around the relatively immobile cations in a way such that pairs of O^{2-}-ions and vacancy defect (O^{2-}–vac pair) are in greater numbers in some cation environments than others. The difference in O^{2-}–vac pair numbers is shown to originate from entropic effects in the cation environments. As a consequence of the entropic effect, the O^{2-}–vac pair distribution is practically independent of temperature between 974 and 1874 K. Even the O^{2-} ion hopping rate and barrier are influenced by the entropic effect. This observation is in stark contrast with the standard belief that energetic interactions dominate in solid oxide materials, and entropic factors can be ignored. By analyzing the cation environment effect using *k*-means clustering, nine environment clusters are identified that are associated with unique values of the activation barrier and entropy. Comparisons to yttria stabilized zirconia (YSZ), yet another popular fast oxygen ion conductor, are made.

197by Development of a Modifiable Atomistic Cellulose Nanocrystal Model

Arash Elahi, Santanu Chaudhuri

The extremely low dispersibility of the carboxylic acid cellulose nanocrystals (CNC-COOH) in non-aqueous organic solvents prevented their massive utilization in many research or industrial areas. For example, despite the high affinity of the CNC-COOHs to stabilize graphene flakes, their low dispersibility in ethanol forbids their integration into graphene-based inks formulated in the ethanol solvent. To the best of our knowledge, the improvement of CNC-COOH's dispersibility in organic solvents using covalent surface modification has not been investigated. Hence, in the current study, we carried out molecular dynamics simulation to investigate the impacts of the surface functionalization on CNC's dispersibility, and compared the results against our experimental measurements. Our potential of mean force (PMF) calculations revealed functionalization of CNC-COOHs with alkyl groups (CNC-COOH-alkyl), e.g., ethyl, butyl, hexyl, and octyl, enhances dispersibility in ethanol compared to the pristine CNC, either due to the reduced strength between the binding partners or a developed kinetic barrier along the binding pathway.

Our experimental dispersibility tests confirmed our simulation predictions, wherein the alkyl chain length increment increased the dispersibility. The well dispersion of CNC-COOH-alkyls in ethanol sets the stage for their integration as stabilizers in ethanol or other organic solvents.

544a Enhanced Sampling Molecular Dynamics Simulations of Polymorphism in Molecular Crystals

Oren Elishav, Roy Podgaetsky, T. Kyle Vanderlick, Barak Hirshberg

Polymorphism affects the properties of molecular solids and is crucial in processing pharmaceuticals, pigments, agrochemicals, and energetic materials. Molecular dynamics (MD) simulations can provide important information about these materials' polymorphic transitions and crystallization processes by estimating the underlying free energy surface (FES). However, constructing the FES from MD simulations is difficult due to the long timescales (>1 μs) associated with crystal phase transitions. Enhanced sampling methods, such as well-tempered Metadynamic (WT-MetaD), bias the system to increase the occurrence of phase transitions during an MD simulation. MetaD requires defining reaction coordinates (collective variables, CVs) that sample the slowest degree of freedom of the process. In conformational polymorphism, the constituent molecules crystallize in diverse forms, which differ by the molecular conformation and not just by the molecular packing or lattice parameters. Thus, finding good CVs to accelerate transitions between the different polymorphs is challenging. Data-driven approaches can be used to catalyze CV discovery.

Herein, we used MD simulations and dimensionality reduction algorithms to describe conformational polymorphic transitions of CL-20, a complex energetic material with three anhydrous conformational polymorphs (β, γ, and ϵ) at ambient pressure.[1] We demonstrate that a local molecular CV obtained from short gas phase simulations of isolated conformers can drive polymorphic phase transitions. We successfully converge the FES of the system and obtain the correct qualitative thermodynamic stability order for the three polymorphs. Moreover, we identified several defect and intermediate forms, which are local minima in FES. We next investigate polymorphism in a phosphorescent organic material by combing descriptors of molecular orientation and global molecular packing. The approach allows for identifying polymorphic transitions into new forms. These findings demonstrate the benefits of combining data-driven approaches with biased MD simulations to describe polymorphism in complex systems and find new strategies to obtain desired molecular polymorphs.

[1] Collective Variables for Conformational Polymorphism in Molecular Crystals, Oren Elishav, Roy Podgaetsky, Olga Meikler, and Barak Hirshberg, The Journal of Physical Chemistry Letters 2023 14 (4), 971-976, DOI: 10.1021/acs.jpclett.2c03491

544b The Role of the Solvent in Crystal Nucleation from Solution: A Novel Approach

E. Daniel Cárdenas-Vásquez, Erik Santiso

Background: The study of crystal nucleation poses major challenges for both experimental and simulation studies. The stochastic nature of nucleation, combined with its uncanny ability to amplify the effect of minute molecular-level details, make it very difficult to achieve consistent experimental results even for simple systems. Molecular simulation studies are also challenging, as nucleation is a rare event with a free energy barrier that includes a substantial entropic contribution. Additionally, some of the common assumptions in molecular simulations, such as periodic boundary conditions (which can result in serious finite size artifacts) and constant number of molecules, can greatly bias the results. All theories of nucleation focus on parameters of the solid phase (like number of molecules in the solid), but there is evidence from computational studies that parameters from the solution phase could be equally important.

Results: We defined the liquid order parameters for four different solvents (acetonitrile, ethanol, methanol and water) containing urea at five different concentrations. The behavior of each solvent in the nucleation process has a direct correlation with its nucleation mechanism (i.e. one- or two-step mechanism) found in previous studies, showing that we could predict the behavior and/or pathway using the solvent physical and thermodynamic properties. Additionally, these results can be extended to other solutes in the drug discovery industry.

Conclusions: We show some of the first studies that uses solvent order parameters to predict the behavior of the nucleation of crystals in solution. Consequently, we could generalize this procedure to other liquid structures, which will enable us to generate order parameters for any solvent of interest.

544c On the Challenge of Sampling Multiple Nucleation Pathways: A Case Study of Heterogeneous Ice Nucleation on FCC (211) Surface

Wanyu Zhao, Tianshu Li

Keyword: nucleation rate, forward flux sampling, co-existing nucleation pathways, two-path model

Heterogeneous ice nucleation plays a significant role in a variety of fields ranging from global climate change to drug production, yet its microscopic mechanisms can be remarkably intricate even on a simple surface. Such complexity poses a challenge in modeling nucleation using advanced sampling methods. In this study, we explore heterogeneous ice nucleation on the FCC (211) surface with a lattice constant of 4.158 Å using a forward flux sampling (FFS) approach[1] based on mW model[2]. We demonstrate that the commonly employed size-based order parameter fails to describe heterogeneous ice nucleation on the FCC (211) surface, which yields an ensemble of phase points that are virtually composed of "disk-like" seeds that cannot grow into ice. To overcome the issue, we develop a new order parameter λ_k by integrating a geometric constraint on ice seed by computing its maximum ratio in the resolved components of the radius of gyration k.

The new order parameter is found to significantly improve the quality of FFS in the current system[3]. With a proper range of geometric constraints, the calculated nucleation rates are found to be nearly invariant with a k_{cut} between 3 and 5, while a small k_{cut} (<2) places a very strong geometric constraint leading to a lower nucleation rate. The quality of the λ_k was further verified by pB histogram analysis at the critical sizes identified at different k_{cut}. Interestingly, structural analysis of nucleation trajectories shows co-existence of two distinct nucleation pathways. One is a primary-prism-planed (PPP) path that was previously identified in direct MD simulations[4,5]. The other pathway, which was absent from direct MD simulations on FCC (211) surface with the lattice constant of 4.158 Å, yields the secondary-prism-planed (SPP) path of Ih. PPP and SPP both lead to the formation of hexagonal ice but with distinct crystalline orientations. The 4-order of magnitude difference in growth probability of PPP and SPP showed that nucleation of ice on the FCC (211) surface is dominated by PPP, which explains the absence of SPP in previous direct simulations[5].

Although the PPP path has a significantly higher nucleation rate than the SPP path, the less-efficient SPP path is found to overwhelm the ensemble of configurations collected at the early milestones of FFS modeling, which contributes to the substantial statistical uncertainty in the FFS rate constant calculations and exhibits an undesirable dependence on the choice of FFS parameters.

To address this sampling issue, we develop a two-path model to understand how the collection of initial sampling at λ_0, for example, initial milestones and the number of initial configurations, influence the reliability of FFS computations when two competing nucleation pathways are involved. The two-path model, which considers the co-existence of a fast and a slow route in one system, provides a comprehensive, quantitative understanding of the dependence of the final rate constant in FFS on the initial weights and the difference in the growth probabilities of both routes. Such dependence highlights the challenge of modeling crystal nucleation in the presence of multiple pathways, particularly in pursuit of a quantitative description of the key rate constant that governs the true nucleation dynamics. Our study also suggests a few general strategies for improving the accuracy of FFS when exploring unknown but complex systems.

Acknowledgement:

The work is supported by NSF through award CBET-2053330.

References:

- J. Allen, D. Frenkel, and P. R. ten Wolde, J. Chem. Phys. 124, 024102 (2006).
- Molinero and E. B. Moore, J. Phys. Chem. B 113, 4008 (2009).
- Zhao and T. Li, J. Chem. Phys. 158, 124501 (2023).
- Fitzner, G. C. Sosso, S. J. Cox, and A. Michaelides, J. Am. Chem. Soc. 137, 13658 (2015).
- Fitzner, G. C. Sosso, F. Pietrucci, S. Pipolo, and A. Michaelides, Nat. Commun. 8, 2257 (2017).

544e Novel Patch Formation in Ligand-Grafted, Anisotropic, Nanoparticle Stencils

Tommy Waltmann, Thi Vo, Ahyoung Kim, Sharon C. Glotzer, Qian Chen

Patch formation on nanoparticle surfaces is a primary determinant of bonding behavior, driving formation of crystal structures with corresponding local valence. Ligand-grafted, anisotropic nanoparticles have previously been found to exhibit patches solely on high curvature regions of the nanoparticle shape[1]. To expand the range of local valences, and therefore range of crystal structures a given nanoparticle shape can assemble, iodide has been utilized in solution to preferentially adhere to specific regions of the nanoparticle surface, preventing patch formation in those locations. We simulate this "atomic stencil" method to demonstrate patch formations in novel locations consistent with experiments. We demonstrate face patch formation in cuboctahedra, rhombic dodecahedra, and a range of previously inaccessible patch morphologies in truncated tetrahedra. These novel patch valences may open doors to the assembly of novel nanoparticle crystal structures.

[1] Kim, A., Vo, T., An, H. et al.Symmetry-breaking in patch formation on triangular gold nanoparticles by asymmetric polymer grafting. Nat Commun13, 6774 (2022).

544f A More Efficient Model for the Emergent Ionic Conduction in Aliovalently Doped Zirconia

Bharathi Bandi

Fast ion conductors are used in solid state devices like fuel cells, electrolyzers and batteries. Despite their growing importance, a connection between microscopic hopping parameters and macroscopic ionic conduction, which is essential for materials design, is often not available. This is particularly true for aliovalently-doped oxide materials. Using yttria stabilized zirconia (YSZ) as a prototypical system, we demonstrate the complexity arising from anisotropic oxygen ordering when ions migrate in response to an electric field. Statistical analysis of classical molecular dynamics trajectories provides atomistically-resolved quantities as a function of the field strength and various local cation arrangements that make up the YSZ structure. The effect of anion ordering on individual oxygen ion hopping rates is determined. Contrary to standard rate theories, the hopping rate constant is found to be decrease with the field strength both in the direction of the applied field and against it. Despite this, the current density that emerges at the macroscopic scales considering contributions from different cation arrangements is consistent with Ohm's law and other experimental observations. These insights suggest that it might be possible to build reduced complexity ionic conduction model for solid-state fast ion conductors.

544g A Probabilistic Microkinetic Modeling Framework for Catalytic Surface Reactions

Aditya Kumar, Abhijit Chatterjee

We present a probabilistic microkinetic modeling (MKM) framework that incorporates the short-ranged order (SRO) evolution for adsorbed species (adspecies) on a catalyst surface. The resulting model consists of a system of ordinary differential equations. Adsorbate–adsorbate interactions, surface diffusion, adsorption, desorption, and catalytic reaction processes are included. Assuming that the adspecies ordering/arrangement is accurately described by the SRO parameters, we employ the reverse Monte Carlo (RMC) method to extract the relevant local environment probability distributions and pass them to the MKM. The reaction kinetics is faithfully captured as accurately as the kinetic Monte Carlo (KMC) method but with a computational time requirement of few seconds on a standard desktop computer. KMC, on the other hand, can require several days for the examples discussed. The framework presented here is expected to provide the basis for wider application of the RMC-MKM approach to problems in computational catalysis, electrocatalysis, and material science.

544h Bridging the Gap between the Experimental and Computational Catalysis Communities: Recent Advances and Challenges

Rachita Rana, Simon Bare, Ambarish Kulkarni

Theory and modeling are often used to elucidate atomistic phenomena that are experimentally inaccessible.[1–3] Although modeling is frequently used to explore catalytic performance of materials for a plethora of reactions, relatively less attention has been focused on using high-throughput simulations to quantitatively interpret experimental measurements.[4] Recently, we have demonstrated one such approach, where a database of structures (obtained from density functional theory (DFT)) is used to identify the catalytic site that is most consistent with spectroscopy (e.g., XAS (x-ray absorption spectroscopy) and FTIR (Fourier transform infrared spectroscopy)) and microscopy data.[5] This approach, denoted as QuantEXAFS is especially powerful for (relatively) well-defined materials like atomically dispersed catalysts. Specifically, by using DFT to interpret experimental data, this approach aims to bridge the gap between the computational catalysis and the catalyst characterization communities.

The overall goal of this talk is to discuss the role (and limitations) of QuantEXAFS within the catalysis science community. Using Pt/MgO calcined at 700 °C and 400 °C as a prototypical example, I will first discuss the limitations of conventional EXAFS analyses that led to the development of QuantEXAFS. Further, I will show how QuantEXAFS enabled the identification of different unique sites for the two calcination temperatures.[5,6] I will show results to demonstrate that the complexity of zeolites can be handled and this method is robust in dealing with different support materials.[7] I will present a summary of all the recent applications of this method on different catalytic materials with both internal and external collaborators. While QuantEXAFS is a useful tool, it has limitations and these will be discussed in the talk along with the best practices and common pitfalls. In the end, I'll talk about unpublished results that show how we can start tackling site heterogeneity– the fractions of different atomically dispersed sites present in a single catalytic sample.

With more and more emphasis on machine learning and FAIR data practices it becomes essential to explore tools like QuantEXAFS. Moreover, it is important to equip the catalysis community with powerful (combined) experimental-theory driven methods. Bridging the gap between experimental catalysis and computational catalysis communities requires participation from many communities, I'll highlight the key outlooks and future directions for such advanced bridging approaches in the field of fundamental catalysis.[8]

References

(1) Li, Y.; Frenkel, A. I. *Acc. Chem. Res.* **2021**, *54* (11), 2660–2669. https://doi.org/10.1021/acs.accounts.1c00180.

(2) Timoshenko, J.; Duan, Z.; Henkelman, G.; Crooks, R. M.; Frenkel, A. I. *Annu. Rev. Anal. Chem.* **2019**, *12*, 501–522. https://doi.org/10.1146/annurev-anchem-061318-114929.

(3) Xia, Y.; Campbell, C. T.; Roldan Cuenya, B.; Mavrikakis, M. *Chem. Rev.* **2021**, *121* (2), 563–566. https://doi.org/10.1021/acs.chemrev.0c01269.

(4) Desai, S. P.; Ye, J.; Zheng, J.; Ferrandon, M. S.; Webber, T. E.; Platero-Prats, A. E.; Duan, J.; Garcia-Holley, P.; Camaioni, D. M.; Chapman, K. W.; Delferro, M.; Farha, O. K.; Fulton, J. L.; Gagliardi, L.; Lercher, J. A.; Penn, R. L.; Stein, A.; Lu, C. C. *J. Am. Chem. Soc.* **2018**, *140* (45), 15309–15318. https://doi.org/10.1021/jacs.8b08550.

(5) Chen, Y.; Rana, R.; Sours, T.; Vila, F. D.; Cao, S.; Blum, T.; Hong, J.; Hoffman, A. S.; Fang, C. Y.; Huang, Z.; Shang, C.; Wang, C.; Zeng, J.; Chi, M.; Kronawitter, C. X.; Bare, S. R.; Gates, B. C.; Kulkarni,

A. R. *J. Am. Chem. Soc.* **2021,** *143* (48), 20144–20156. https://doi.org/10.1021/jacs.1c07116.

(6) Chen, Y.; Rana, R.; Huang, Z.; Vila, F. D.; Sours, T.; Perez-Aguilar, J. E.; Zhao, X.; Hong, J.; Ho, A. S.; Li, X.; Shang, C.; Blum, T.; Zeng, J.; Chi, M.; Salmeron, M.; Kronawitter, C. X.; Bare, S. R.; Kulkarni, A. R.; Gates, B. C. *J. Phys. Chem. Lett.* **2022**. https://doi.org/10.1021/acs.jpclett.2c00667.

(7) Felvey, N.; Guo, J.; Rana, R.; Xu, L.; Bare, S. R.; Gates, B. C.; Katz, A.; Kulkarni, A. R.; Runnebaum, R. C.; Kronawitter, C. X. *J. Am. Chem. Soc.* **2022**, *144* (30), 13874−13887. https://doi.org/10.1021/jacs.2c05386.

(8) Rana, R.; Vila, F. D.; Kulkarni, A. R.; Bare, S. R. *ACS Catal.* **2022,** (12), 13813−13830. https://doi.org/10.1021/acscatal.2c03863.

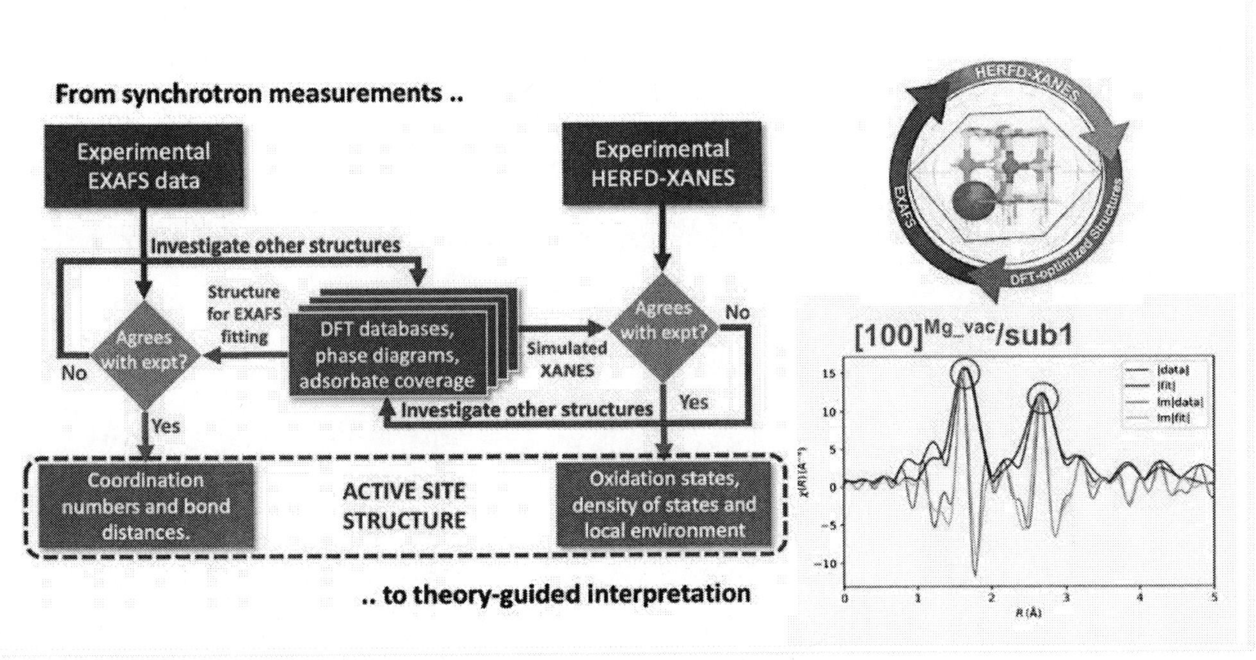

544i Dynamic Molecular Modes to Interpret NMR Relaxation at High Frequencies: Theory and Applications

Thiago Jose Pinheiro Dos Santos, Dilip Asthagiri, Philip Singer, Walter Chapman

Nuclear magnetic resonance (NMR) relaxation is a powerful technique for probing matter non-destructively, with applications spanning medicine to material science to petrophysics. At high NMR frequencies, quantum effects are usually negligible, and a semi-classical statistical mechanical description is sufficient to model relaxation. Traditional theories of NMR relaxation assume the relaxation autocorrelation function is mono-exponential, but calculated relaxation evinces a rich, multi-exponential behavior. In this work, we develop an approach to model and interpret this multi-exponential behavior. Drawing inspiration from the idea of inherent structures in liquids, we find that there are dynamic molecular modes that the system samples, with each mode following a mono-exponential decay. In this talk we will present this idea using simple, physical models within a rigorous statistical mechanical development. The key point is that based on the dynamic molecular modes, our statistical mechanical description can predict such multi-exponential decay accurately. These molecular modes contain important information about the structure/dynamics of the system.

544j Development of Heteroatomic Constant Potential Method for Mxene-Based Supercapacitors

Xiaobo Lin, Shern R. Tee, Paul R.C. Kent, Debra J. Searles, Peter T. Cummings

Molecular dynamics (MD) simulations using the constant potential method (CPM) have been proven to be an essential technique for studying the charge storage and charging dynamics of homogeneous electrodes.[1–3] However, the CPM cannot capture the distinct atomic nature of heteroatomic electrodes, such as MXenes.[3] To overcome this limitation, we developed the heteroatomic constant potential method (HCPM) for MXene supercapacitors. The proposed HCPM model takes into account the electronegativities of various atoms in heteroatomic electrodes. To predict the charge responses of MXene atoms, this model is adjusted to align with the density functional theory (DFT) results by using derivative-free optimization. We also performed molecular dynamics simulations using both HCPM and CPM for MXene electrodes with solvent-in-salt electrolytes. Although the two methods show similar accumulated charge storage on the electrodes, the results showed that HCPM provides a more reliable depiction of electrode atom charge distribution and charge response compared with CPM. Furthermore, the simulations showed that the two methods exhibit similar anion structures near the MXene surface, while cations are drawn closer to the MXene surface in HCPM. These results highlight the crucial influence of electrode electronegativity and charge response on the atom charge distribution on electrodes and electrolyte structure. In the future, the tunable parameters of our HCPM model can be easily adapted to other heteroatomic electrodes, including different types of MXenes, metal-organic frameworks (MOFs), and carbonaceous electrodes with structural defects, providing new avenues for exploring heteroatomic supercapacitors.

Reference

(1) Reed, S. K.; Lanning, O. J.; Madden, P. A. Electrochemical Interface between an Ionic Liquid and a Model Metallic Electrode. *Journal of Chemical Physics* **2007**, *126* (8).

(2) Merlet, C.; Rotenberg, B.; Madden, P. A.; Taberna, P. L.; Simon, P.; Gogotsi, Y.; Salanne, M. On the Molecular Origin of Supercapacitance in Nanoporous Carbon Electrodes. *Nat Mater* **2012**, *11* (4), 306–310.

(3) Xu, K.; Shao, H.; Lin, Z.; Merlet, C.; Feng, G.; Zhu, J.; Simon, P. Computational Insights into Charge Storage Mechanisms of Supercapacitors. *ENERGY & ENVIRONMENTAL MATERIALS* **2020**, *3* (3), 235–246.

544k Estimating and Correcting DFT Error of Metal-Organic Frameworks through Molecular Derivatives

Yeongsu Cho, Aditya Nandy, Chenru Duan, Heather Kulik

Metal-organic frameworks (MOFs) are composed of metal clusters linked by organic molecules, forming a porous structure that is suitable for a range of applications, including catalysis, hydrogen storage, and CO_2 capture. While an accurate theoretical analysis is crucial for characterizing and designing functional MOFs, the presence of transition metals makes MOFs exhibit strong static correlation and considerable errors in their density functional theory (DFT) calculations. Additionally, post-DFT methods are computationally prohibitive for MOFs due to the large number of atoms per unit cell. We estimate the magnitude of static correlation of MOFs using low-cost multireference diagnostics based on finite-temperature DFT, which is the first time they are applied to solid-state systems. By comparing the behavior of the diagnostics on MOFs and their molecular derivatives, we demonstrate that the DFT-based multireference diagnostics are equally applicable to solids as to their molecular derivatives. Moreover, we show that estimates provided by multireference diagnostics of a MOF have a good correlation with those of its molecular derivatives, which can be calculated much more affordably in comparison to those of the full MOF. The additivity of the multireference character discussed here suggests the set of molecular derivatives to be a good representation of a MOF, allowing it to be utilized to extract empirical parameters. Therefore, we propose to determine the Hubbard U parameter by fitting DFT results of molecular derivatives to their theoretical benchmark, and apply the same parameter to DFT+U calculations of MOFs to effectively correct the multireference error. The quality of fitted U parameters was evaluated on the adsorption energy of N_2, O_2, and CO_2 gas molecules. The ability to perform high-level calculations for molecular systems and the low cost of DFT+U make this procedure a practical strategy for designing MOFs with improved accuracy and efficiency.

544l Optimization of Vaspsol Solvation Free Energy Predictions

Eric Fonseca, Sean Florez, Richard G. Hennig

Density functional theory (DFT) can accurately predict material properties and reaction barriers. However, DFT is often limited to small system sizes due to high computational costs. In addition, many properties and reaction barriers are dramatically different when solvent molecules surround the atomic system. To approximate the solvation effect, computational chemists use continuum models to mimic the countless number of solvent molecules in these systems. Continuum models attempt to capture the effect of the solvent on solute molecules and surfaces while dramatically reducing the computational cost. VASPsol uses a polarizable continuum model within VASP, a plane-wave DFT code. VASPsol has three parameters that control the cavity shape and, thus the solvation energy. These three parameters are the electronic cutoff of the cavity () ,the breadth of the diffuse cavity boundary (), and the cavity surface tension (). The present work sought to optimize the VASPsol cavity parameters by minimizing solvation energy errors compared to experimentally measured values.

Our work will demonstrate the techniques used to reduce VASPsol's prediction error of solvation free energies against a dataset of neutral compounds solvated in various solvents. We apply three methods to find a set of VASPsol cavity parameters that result in lower errors in the entire Truhlar set. First, we attempt a grid search across electronic cutoff and transition of the diffuse cavity. The surface tension of the cavity is initially fixed for the grid search. We also apply a Nelder-Mead search for the absolute minimum error for all three VASPsol parameters, including the surface tension.

For our machine learning descriptors, we used the COSMO-SAC sigma profile. The sigma profile is a histogram of the charge density on the solvent-accessible surface area for each molecule as defined by the COSMO procedure. The COSMO sigma profile is thus used as a feature in the general analysis of the molecular set. Using this constructed dataset of sigma profiles, VASPsol parameters (and), and solvation energy errors, we can construct artificial neural networks trained to learn the errors that VASPsol produces for a given molecule and VASPsol parameters. We used experimental solvation energies from the Truhlar Minnesota dataset. We first assemble a set of 9 candidate molecules balanced by using sigma profiles. We access the sigma profile of these molecules using online databases, such as the Virginia Tech dataset by Mullins et al. Then we build a set of 9 training molecules from chemical intuition. These molecules also have an average sigma profile that matches the Truhlar set average sigma profile.

Parameters obtained from grid searches or direct optimization with the training set produce larger errors against the entire Truhlar set. Despite our efforts, the training set appears biased and caused us to overfit our VASPsol parameters. From each procedure, the minima and their corresponding errors can be found in figure 1. We can see that the traditional approaches with a small dataset resulted in worse errors than default VASPsol parameters. As a new approach, we train an artificial neural network on our dataset constructed from benchmarks and prior optimization attempts. This artificial neural network will be used to guide the optimization towards a minimum for the entire Truhlar dataset when computed with VASPsol. We train the artificial neural networks with the grid search, Nelder-Mead optimizations, and other validations conducted with the VASPsol code against the Truhlar set. The neural network then predicts the errors VASPsol produces for each molecule given the sigma profile representation and the VASPsol cavity parameters. This approach, known as delta-learning, will allow us to minimize the outputs of our neural network by changing the VASPsol cavity parameters. The extended benefit of this method is that we obtain a model that generalizes how VASPsol predictions deviate in response to parameter changes for any molecule. By analyzing the resultant errors across chemical groups, we optimize VASPsol parameters for multiple solvents. We show that VASPsol can lead to more accurate simulations for the larger community using these optimized parameters.

Method	n_c [e/Å^3]	σ	τ [eV/Å^2]	MAE [kcal/mol]
Default VASPsol	0.00250	0.60	5.25E-4	1.17
Nelder-Mead/VASPsol	0.00275	0.66	4.72E-4	1.48
Nelder-Mead/ML	**0.00366**	**0.58**	**9.36E-4**	**1.10**

+6% Improved ΔG_{solv}

346b The Effect of Pressure and Salinity on Gas/Brine Interfacial Tension for Equimolar Mixtures of H_2-CH_4 and H_2-CO_2

Salem Alshammari, Safwat Abdel-Azeim, Abdulaziz Alqasim, Moataz Abu Alsaud

The effect of salinity, pressure, and different cushion gases like carbon dioxide and methane on the interfacial tension (IFT) between H_2 and brine is essential in many applications like geological hydrogen storage. We have used molecular dynamics simulation to calculate the surface tension of H_2-CH_4/brine and H_2-CO_2/brine systems under varying pressures and salinities. In preparing the simulation box, the number of molecular units for the gas phase was calculated using the Peng-Robinson equation of state with simple van der Waals mixing rules. The simulation box ($L_x = 4$ nm, $L_y = 4$ nm, $L_z = 12$ nm) was built; energy minimization was carried out using the steepest decent algorithm until the maximum force was converged to 100 kJ/mol. Subsequently, 0.1 ns NVT equilibration and 50 ns production runs were performed. The cut-off distance for van der Waals interactions was set to 1.2 nm while electrostatic interactions were treated using the particle mesh Ewald summation.

The simulation trajectories were analyzed to construct density profiles and use the principal components of the pressure tensor in order to calculate the IFT. It was found that higher ionic strength resulted in greater surface tension which is a consequence of the perturbation of water structure brought about by the ions. Furthermore, the H_2-CO_2 mixture had a lower IFT compared to H_2-CH_4. The surface tension manifests due to the difference in net forces between molecules in the bulk liquid solution and those near the interface. Because the CO_2 mixture interacts with the liquid more strongly, the surface tension will be less than that of the CH_4 mixture. This was verified by calculating the interaction energies between the phases, and the resulting trend matched the IFT one.

346c Molecular Simulations of H_2 and CO_2 Intercalation into Montmorillonite Interlayers in a Geological Gas Storage

Tuan Ho, Yifeng Wang, Carlos F. Jove-Colon

CO_2 capture and geological sequestration have been considered important strategies to reduce the greenhouse gas effect. Meanwhile, H_2 produced from water using renewable energy and stored underground will be important components in a H_2 supply chain in a future H_2-based economy. When injected into a reservoir, gas interacts with earth materials that might impact the safety and economic feasibility of a storage. In this talk, we will discuss gas intercalation into smectite that is present in caprock. Intercalation traps CO_2, thus improving CO_2 storage integrity. However, intercalation causes H_2 loss. Using molecular dynamics (MD) simulation, we calculate free energy landscapes of gas intercalation into montmorillonite interlayers as a function of structural charge distributions and interlayer hydration states. Our results reveal that hydrated clay interlayers promote gas uptake into the hydrophobic (e.g., siloxane groups) domain, while inhibiting gas sorption near the hydrophilic sites (e.g., charge sites). On average, the CO_2/water ratio in interlayers is higher than CO_2/water ratio in bulk water. However, H_2/water ratio in interlayers is comparable or less than H_2/water ratio in bulk water. Compared to CO_2, the amount of H_2 intercalated into hydrated interlayers is about one to two orders of magnitude smaller. Our results qualitatively predict that H_2 loss due to intercalation into clay-rich caprock and leak through clay interlayers, if any, is limited.

Sandia National Laboratories is a multimission laboratory managed and operated by National Technology & Engineering Solutions of Sandia, LLC, a wholly owned subsidiary of Honeywell International Inc., for the U.S. Department of Energy's National Nuclear Security Administration under contract DE-NA0003525.

346d Mechanistic Study of Reaction and Transport Behaviors in Hybrid CO_2 Capture Materials By Multi-Scale Modeling

Xiaoyang Shi, Ga Hyun Lee, Ah-Hyung Park

The rational discovery of novel dual-functional materials with desired properties is essential to the advancements of the combination of CO_2 capture and conversion. Recent innovations in computational chemistry, data informatics, and machine learning can rationally guide the material design and will lead to faster kinetics in the development of energy-related industries. In this study, we discovered the underlying mechanism of dual-functional materials, such as nanoparticle organic hybrid materials (NOHMs) of MOFs, during CO_2 capture with or without the existence of water molecules. We studied molecule transport and diffusion behaviors in dual-functional materials. Due to the probable existence of nonselective voids or other defects, it has been hypothesized that the gas diffusion behavior at the interface between the polymer matrix and functional particles would have a significant impact on the global performance of composites. We also employed Molecular Dynamics to provide an unprecedented microscopic image of the transport of CO_2, H_2O, and O_2 as a mixture in all regions of dual-functional materials.

346e Computational Discovery and Evaluation of Plastic-Binding Peptides for Microplastic Remediation

Michael Bergman, Carol Hall, Xingqing Xiao

Plastic pollution on the micrometer or nanometer scale, termed microplastics or nanoplastics (MNPs), is a pressing environmental issue. There are millions of tons of MNPs in just the top 200 meters of the Atlantic Ocean, and MNPs are also found in the arctic ice caps, the atmosphere, and many food and water sources. The ubiquitous presence of MNPs causes many organisms, including humans, to consume MNPs daily, which is believed to have negative health impacts. The environmental and health risks associated with MNPs have spurred much research into creating methods that detect and remove MNP pollution from the environment.

We aim to computationally design plastic-binding peptides to help remediate MNP pollution. Specifically, we are working with collaborators to remediate MNP pollution in bodies of water by conjugating the designed peptides to active particle microcleaners to help capture MNPs, incorporating the peptides into liquid crystal sensors to create a platform for detecting and characterizing MNP waste, and expressing the peptides on the surface of engineered microbes to increase cell adhesion and subsequent degradation of MNPs.

In this presentation, we will discuss our efforts to design peptides that bind to four of the most common plastics: polyethylene, polypropylene, polystyrene, and PET. Towards this goal, we first generated models of plastic surfaces via molecular dynamics simulations, then developed a score function that captures the key intermolecular interactions between the peptide and the plastic surface. Next, we developed a procedure that produces an ensemble of adsorbed conformations to sample the vast conformational space available to peptides in the adsorbed state. We then use **Pep**tide **B**inding **D**esign (PepBD), a Monte Carlo algorithm paired with simulated annealing, to optimize the peptide sequence and structure for each starting conformation. The top-scoring peptides from the ensemble of starting states were tested using steered molecular dynamics simulations to obtain approximate adsorption free energies. The best peptides from simulations were tested experimentally through atomic force microscopy and biolayer interferometry. Our results indicate the peptides designed by PepBD have higher affinity for plastics than the peptides previously found using phage display. Lastly, we also investigate the properties that endow a peptide with high affinity for plastics, and how these properties vary for different plastic targets.

346f Experimental and Computational Evaluation of the Interaction between per- and Polyfluoroalkyl Substances with Microplastics in Aqueous Systems

Gabriel Barbosa, Jasneet Pala, Medha Kasula, Milad Esfahani, C. Heath Turner

Per- and polyfluoroalkyl substances (PFAS) are surface-active compounds composed of alkyl fluorinated hydrophobic tails, applied as industrial surfactants mainly due to their chemical stability. The development of health problems like diabetes, high cholesterol, cardiovascular disease, thyroid, kidney, and immune system dysfunctions have been recently linked to long-term human exposure to PFAS, such as perfluorooctane sulfonate (PFOS). In order to develop a better understanding of the fate of these molecules in aqueous systems, it is important to quantify the interaction of these molecules with different microplastic species. Here, we experimentally assess the adsorption behavior of PFOS in aqueous media onto the surfaces of polyethylene (PE), polypropylene (PP), and polyvinyl chloride (PVC). The experimental behavior is further analyzed using molecular dynamics simulations to calculate the PFAS superficial properties, such as the PFAS-PFAS virial coefficient and transfer free energy, in which PFAS molecules with different alkyl sizes are evaluated. The adsorption mechanism of four different perfluorooctyl PFAS molecules on PE, PP, and PVC nanoplastic particles is evaluated by inspecting the potential-of-mean-force of pulling the PFAS molecules from the aqueous bulk phase to the plastic-water interfacial region. The molecular simulation results suggest a pronounced dispersive interaction between the polymer and PFAS molecules for PE and PP; while the dispersive interaction dominates the PVC systems, as compared to PE and PP, a much stronger coulombic PFAS-PVC interaction is found. Overall, evaluating the relevant molecular features involved in the adsorption of aqueous phase dispersed PFAS molecules, as performed in this work, is essential in understanding the lifecycle of PFAS transport and designing novel polymetric adsorbent materials.

346g Differences in the Performance Characteristics of Dye-Sensitized Solar Cells Using Aqueous and Organic Electrolytes: Insights from Atomistic and Molecular Simulations

Victor Fabiyi, Eunsu Paek, Sitaraman Krishnan

Visible-light-absorbing dyes adsorbed on broad-band gap oxide semiconductors such as titania have shown remarkable power conversion efficiencies when used as photoanodes in dye-sensitized solar cells (DSSCs). The dye molecules oxidize upon illumination, transferring the electrons to the titania particles. A redox electrolyte, such as an acetonitrile solution of iodide and triiodide, is required to regenerate the oxidized dye and maintain continuity in the photovoltaic activity. DSSC-based photoelectrochemical cells are being explored to desalinate brackish water and produce potable water [1]. Applications such as these preclude the use of toxic and volatile solvents in the preparation of the redox electrolyte. Water is the preferred solvent. However, DSSCs show a significant performance loss when water is used instead of acetonitrile as the electrolyte solvent.

This work used atomistic and molecular simulations of selected dyes (known to exhibit photovoltaic activity in experiments) to gain fundamental insights into the differences between the aqueous and organic electrolyte interfaces. Critical optical and electronic properties of the dyes were predicted using density functional theory (DFT) calculations. Also, a molecular dynamics (MD) model was developed to study the behavior of the dye/titania interface in water and acetonitrile. Based on the DFT calculations, the absorption spectra and electronic properties of the implicitly solvated dyes were similar in the two solvents. However, MD simulations presented some fundamental differences between the dye exposed to water vs. acetonitrile interfaces. These include differences in the structural re-orientation of the dye moieties, the solvation behavior of the dye functional groups, and the transport properties of the electrolyte ions. The computational framework described in this work would be helpful in identifying ideal materials for the DSSC and rationally designing the electrode-electrolyte interface using molecular predictions.

Reference:

[1] Mohandass, G.; Kim, T.; Krishnan, S. Continuous Solar Desalination of Brackish Water via a Monolithically Integrated Redox Flow Device. *ACS ES&T Engineering* **2021**, *1*, 1678−1687. https://doi.org/10.1021/acsestengg.1c00266

346h Evaluating Atomical Level Interfacial Structures of Multifunctional Energy Minimizing Surfaces Using Molecular Dynamics Simulations

Nicholas C. Craven, Allison Cordova-Huaman, Chris R. Iacovella, G. Kane Jennings, Clare McCabe

Monolayer films are well recognized as a tool to design surfaces with tuned surface interactions (e.g. water- or oil-repellent surfaces) through modification of the film chemistry[1]. Such films have also been used to design films that are tunably responsive to external media, such as different solvents or pH[2,3]. However, these films are typically prohibitively slow in their responsiveness. In this work we aim to design surfaces that are immediately and reversibly switchable in response to a change in solvent through molecular dynamics simulations. Atomistic amorphous silica surfaces have been functionalized with short alkylsilane chains terminated with functional groups that are composed of competing hydrophobic and hydrophilic subgroups using the MoSDeF design framework[4]. Such terminal groups can achieve a dynamic switching response through molecular reorientation of the individual hydrophobic/hydrophilic groups in response to switching from a hydrophilic to a hydrophobic solvent. The relative size of these groups are varied and the difference in film structure is studied under water and n-hexane. It is shown that the interface is responsive to the nature of the solvent, and the degree of switchability is strongly affected by the length of the terminal group. Parameters are evaluated as quantifiable metrics for switchability which provide information about how the terminal group orients to define the solid liquid interface. These novel parameters will act as an objective function for understanding the types of responsive switching that can occur within the vast parameter space of these films.

References

1. M. C. Stuart *et al.*, "Emerging applications of stimuli-responsive polymer materials," *Nature Materials*, **2010**, 9: 101-113.
2. S. M. Kang, I. You, W. K. Cho, H. K. Shon, T. G. Lee, I. S. Choi, J. M. Karp, H. Lee, "One-Step Modification of Superhydrophobic Surfaces by a Mussel-Inspired Polymer Coating" *Int Ed Engl.* **2010**, 49(49): 9401–9404.
3. Bo He and Junghoon Lee, 'Dynamic wettability switching by surface roughness effect,' The Sixteenth Annual International Conference on Micro Electro Mechanical Systems, **2003**, MEMS-03, pp. 120-123.
4. A. Z. Summers, J. Gilmer, C. Iacovella, P. Cummings, C. McCabe, "MoSDeF, a Python Framework Enabling Large-Scale Computational Screening of Soft Matter: Application to Chemistry-Property Relationships in Lubricating Monolayer Films," *JCTC.*, **2020**, 16: 1779-1793.

346i Surface Modification of Cellulose Nanocrystals Using Alkyl Groups

Arash Elahi, Santanu Chaudhuri

The extremely low solubility of the cellulose nanocrystals (CNCs) in non-aqueous systems prohibited the integration of this novel compound into many research or industrial areas. For example, despite the high affinity of the CNCs to stabilize graphene flakes, their insolubility in ethanol forbids their utilization in graphene-based inks formulated in the ethanol solvent. Also, to the best of our knowledge, there has been a rare number of attempts to improve the CNCs' dispersibility in ethanol by surface modification. Hence, in the current study, we carried out molecular dynamics simulation to investigate the fundamental reasons for the CNCs' low dispersibility in ethanol and the impacts of the surface functionalization on their phase behavior. The solvation free energy calculations of a single cellulose chain in ethanol, as well as the end-to-end distance measurements, revealed that the ethanol molecules dissolve the cellulose chains, and the solute molecule preserves its linear form in the ethanol maintaining its maximum interaction with the solvent. However, the umbrella sampling simulations and the gained potential of mean force (PMF) profile explained that the aggregation of the CNCs through their hydrophilic surfaces drives the CNCs' insolubility in ethanol. The PMFs derived from the binding of functionalized CNCs (CNC-CO-Rs) with alkylamine groups suggested enhanced solubility in ethanol compared to the pristine CNCs, either due to the reduced strength between the binding partners or a developed kinetic barrier along the binding pathway.

346j Effect of Cations on the Adsorption of NO_2^-/NO_3^- Corrosion Inhibitors for STEEL Reinforced Concrete

Ahmed Mohamed, Donald Visco Jr., David Bastidas

To enhance the tensile strength and extend the lifespan of concrete structures, carbon steel rebars are embedded in concrete. In this environment, rebars are naturally protected from corrosion, due to the high alkalinity of the concrete environment (pH ~12.6) creating a protective passive layer. Nonetheless, external environmental factors such Cl^- attacks caused from marine proximity or deicing salts can promote the passivity breakdown. Cl^- ions ingress from the surface of the concrete to the rebar and once it exceeds a certain concentration threshold, then an autocatalytic acid hydrolysis reaction will occur lowering the pH locally and causing the breakdown of the passive – initiating corrosion. Accordingly, expansive corrosion products will accumulate, increasing the pressure in the system and compromising the integrity of the structure. The most common proactive corrosion prevention method used in concrete are corrosion inhibitors as they are economic, effective, and easy to apply. Nitrites (NO_2^-) and nitrates (NO_3^-) are among the most commonly used and effective corrosion inhibitors for this specific environment. There are many studies found in literature associated with NO_2^-/NO_3^- based corrosion inhibitors in different media. However, all these studies focused on one salt form of these inhibitors, either Ca^{2+} or Na^+ at a constant temperature.

To this end, the effect of the cationic species ionically bounded to NO_2^-/NO_3^- corrosion inhibitors was studied for carbon steel rebars in 0.6 M Cl^- simulated concrete pore solution. A 1:1 molar ratio of inhibitor to Cl^- was applied, and the corrosion inhibition efficiency of $NaNO_2$, $Ca(NO_2)_2$, and $Ca(NO_3)_2$, and NaNO3 were evaluated using potentiodynamic polarization (PDP) and electrochemical impedance spectroscopy (EIS). Different temperatures were used (25, 35, 45, and 55 °C) to study the effect of temperature on inhibition and find the activation energy of the corrosion process in the absence and presence of the different inhibitors tested.

According to electrochemical testing NO_2^-/NO_3^- were able to impart protection by forming a stable passive oxide layer achieving an inhibition efficiency (*IE*) up to 85%. The inhibition mechanism of NO_2^- corrosion inhibitors arise from its ability to compete with Cl^- and adsorb though the nitrogen lone pair on the surface of the rebar, oxidizing ferrous ions to form a stable protective layer made of maghemite (Fe_2O_3) and lepidocrocite (γ–FeOOH). The *IE* decreased with increasing temperature, which can be attributed to increased corrosion kinetics or desorption of the corrosion inhibitor. According to the EIS and PDP the best corrosion inhibition at every temperature was achieved by $NaNO_2$ followed by $Ca(NO_2)_2$, $NaNO_3$, $Ca(NO_3)_2$, respectively. The lower *IE* achieved by NO_3^- can be attributed to its inhibition mechanism, since NO_3^- gets reduced to NO_2^-.

The activation thermodynamics of the corrosion process was studied for each inhibitor through an Arrhenius relationship. The activation energy of the corrosion process was 41.15, 35.75, 31.46, 28.28, and 25.35 kJ/mol for $NaNO_2$, $Ca(NO_2)_2$, and $Ca(NO_3)_2$, $NaNO_3$, and blank (i.e. no inhibitor), respectively. This illustrates that the presence of NO_2^-/NO_3^- corrosion inhibitors were able to hinder the initiation of the corrosion reaction. This trend agrees with the electrochemical testing concluding that the best corrosion inhibition is achieved by $NaNO_2$ followed by $Ca(NO_2)_2$, and $Ca(NO_3)_2$, $NaNO_3$. Accordingly, the thermodynamic equilibrium states of the electrolyte was studied since the Na^+ salts of NO_2^-/NO_3^- are performing better than the Ca^{2+} salts in inhibiting the surface of the rebar, although the concentration of the active part (NO_2^-/NO_3^-) of the corrosion inhibitor is constant.

For the purpose of this work, the Pitzer model was utilized to study the equilibrium state of the solutions, as it can accurately model systems with ionic strength under 6 M, making it fit for this study. The activity coefficients were found using PHREEQC – a geochemical modeling open-source code used to calculate different geochemical properties for highly concentrated electrolytes. It was found that Ca^{2+} ions increased the ionic strength, increasing the system's stability, and inducing more ionic interactions compared to Na^+, because of its multivalent charge nature. The increased ionic interactions hinder the mobility of the NO_2^-/NO_3^- ion to inhibit the carbon steel rebar. The presence of Ca^{2+} ions relative to Na^+ decreased the activity coefficient of NO_3^- from 0.46 to 0.42, indicating that there is more active NO_3^- present in Na^+ form than Ca^{2+}, explaining the difference in performance.

474a Predicting Activation Pathway of Integrin Using Generative Adversarial Networks and Targeted Molecular Dynamics Simulations

Siva Dasetty, Tamara C. Bidone, Gregory A. Voth, Andrew Ferguson

Integrins are transmembrane proteins that play a critical role in cell adhesion and cell signaling.[1,2] Upon activation, integrins undergo a large-scale conformational transition from a bent to an extended state. Impaired conformational transitions of integrin are linked to various diseases such as bleeding disorders, immunodeficiency, and cancer.[3] To this end, understanding integrin activation is of fundamental interest and is critical to the development of therapeutics targeting integrin.[1-5]

While crystal structures of bent conformations of different integrins are well-characterized, only fractions of extended conformations are available. Additionally, there is a limited understanding of intermediate conformations of integrin using experimental methods because of their transient nature. Alternatively, studying integrin activation using conventional molecular dynamics simulations require inordinate computational resources because integrins are large heterodimer proteins comprising ~1770 residues. Our goal is to address this problem in characterizing unknown intermediate conformations of integrin by combining generative adversarial networks (GAN)[6] and targeted molecular dynamics simulations.

We demonstrate our approach by training GAN using coarse-grained representations of $\alpha_{IIb}\beta_3$ integrin in an active, two partially inactive and an inactivate state, which are sampled via unbiased all-atom molecular dynamics simulations.[4] We utilize adaptive diffusion maps[7] to learn the latent space of the integrin. Targeted molecular dynamics simulations are employed to predict the all-atom conformations from the coarse-grained transition state structures generated by GAN. Thereby, allowing access to previously unknown all-atom conformations along the activation pathway of integrin. In our talk, we will present our approach and the predicted all-atom conformations of $\alpha_{IIb}\beta_3$ integrin in the transition region and discuss their biophysical validity and significance.

References

[1] Bidone TC, Polley A, Jin J, Driscoll T, Iwamoto DV, Calderwood DA, Schwartz MA, Voth GA. Coarse-grained simulation of full-length integrin activation. Biophysical journal. 2019 Mar 19;116(6):1000-10.

[2] Banno A, Ginsberg MH. Integrin activation. Biochemical Society Transactions. 2008 Apr 1;36(2):229-34.

[3] Zhang Y, Wang H. Integrin signalling and function in immune cells. Immunology. 2012 Apr;135(4):268-75.

[4] Goodman SL, Picard M. Integrins as therapeutic targets. Trends in pharmacological sciences. 2012 Jul 1;33(7):405-12.

[5] Tong D, Soley N, Kolasangiani R, Schwartz MA, Bidone TC. αIIbβ3 integrin intermediates: from molecular dynamics to adhesion assembly. Biophysical Journal. 2022 Dec 23.

[6] Sidky H, Chen W, Ferguson AL. Molecular latent space simulators. Chemical Science. 2020;11(35):9459-67.

[7] Wang J, Gayatri MA, Ferguson AL. Mesoscale simulation and machine learning of asphaltene aggregation phase behavior and molecular assembly landscapes. The Journal of Physical Chemistry B. 2017 May 11;121(18):4923-44.

474b Using Deep Learning to Accelerate Molecular Simulations of RNA Folding

Ayush Gupta, Heng Ma, Arvind Ramanathan, Gul Zerze

The in-vitro experiments studying RNA folding have reported the presence of non-native low-free energy structures in the folding pathways[1]. One bottleneck in using conventional molecular dynamics (MD) simulations in sampling the folding landscapes is that the simulation tends to get stuck in metastable states, as the non-native structures have free energies similar to the folded native structure. To address this, advanced sampling methods can be employed to accelerate the simulation. Many of those methods require defining a set of reaction coordinates or collective variables (CV) i.e. low-dimensional variables that can be used to differentiate between the different metastable states visited by the simulation. Moreover, the success of the sampling often depends on the quality of CVs (i.e. their ability to capture the slowest modes of the process). Obtaining such high quality CVs, however, is a cumbersome task. Neural networks are widely being used to obtain a non-linear set of CVs that can capture the slowest modes of a process. But the current approaches require the availability of a training data set from a reference biased or unbiased simulation to determine the CVs. To circumvent this requirement, we have adapted a deep learning-driven adaptive MD simulation package, called DeepDrive MD that finds undersampled states from an ensemble of running simulations and guides new simulations to visit the unexplored regions of the phase space. For this, we use a convolutional variational autoencoder (CVAE) to represent the visited states in a low-dimensional latent space[2]. Outliers from this learned latent space are identified and treated as starting points for new simulations, resulting in the exploration of newer regions of the phase space. We demonstrate the efficiency of our approach in simulating the folding pathways of a very simple RNA tetraloop GGCGAGAGCC. Our results show that the CVAE is able to learn the slowest degrees of freedom involved in the folding process and successfully samples the transition pathways from the unfolded to the folded state.

References:

1. Zerze, G. H., Piaggi, P. M., & Debenedetti, P. G. (2021). A Computational Study of RNA Tetraloop Thermodynamics, Including Misfolded States. *The Journal of Physical Chemistry B*, *125*(50), 13685-13695.

2. Brace, A., Yakushin, I., Ma, H., Trifan, A., Munson, T., Foster, I., ... & Jha, S. (2022, May). Coupling streaming ai and hpc ensembles to achieve 100–1000× faster biomolecular simulations. In *2022 IEEE International Parallel and Distributed Processing Symposium (IPDPS)* (pp. 806-816). IEEE.

474c Machine Learning-Assisted Design of Deep Eutectic Solvents Based on Uncovered Hydrogen Bond Patterns

Usman Abbas, Qing Shao, Yuxuan Zhang, Joseph Tapia, Jian Shi, Jin Chen

Non-ionic deep eutectic solvents (DESs) have emerged as designer solvents for applications such as catalysis, extraction, and carbon capture and conversion. A major challenge is the lack of an efficient tool to discover DES candidates. Currently, the search relies heavily on the researchers intuition or a trial-and-error process, which leads to a low success rate or bypassing of promising candidates. Recognizing the central role that hydrogen bonds play in the DES formation, this work aims to decipher the hydrogen bond features for DESs and develop machine learning models to predict the potential of a system to be DES based on the hydrogen bond-based descriptors. We first analyze the hydrogen bond properties of 38 known DES and 111 known non-DES systems using their molecular dynamics simulation trajectories. The analysis reveals two features for DES compared to non-DES: (a) the imbalance between the numbers of the two intra-component hydrogen bonds, and (b) more and stronger inter-component hydrogen bonds. Based on the analysis results, we developed 30 machine learning models using ten algorithms and three types of hydrogen bond-based descriptors. The model performance is first benchmarked using their average and minimal ROC-AUC values. We also analyze the importance of individual features in the models and the results are consistent with the simulation-based statistical analysis. Finally, we validate the prediction ability of the models using the experimental results of 34 systems. The Extra Trees outperforms the others in the validation with an ROC-AUC of 0.88. Our work iterates the importance of hydrogen bond in DES formation and shows the potential of machine learning in discovering new DESs.

474d High-Throughput Experimentation and Compositional Screening of Polymer-Based Semiconductor/Insulator Blends for Organic Device Applications

Rahul Venkatesh, Aaron Liu, Haoqun Zhao, Martha Grover, Elsa Reichmanis, Carson Meredith

Conjugated polymer semiconductors are promising materials for their imminent potential in enabling stretchable, flexible, and deformable device applications, such as in transistors, photovoltaics, and others. Recently, polymer semiconductor/insulator blends have gained popularity as a formulation method to afford mechanical properties without sacrificing device performance. For example, when insulating polymers are blended with the active semiconducting layer of organic field-effect transistors (OFETs), the charge-carrier mobility (the main device metric) can match or even exceed devices made with only the single semiconducting component, even at very low compositions of the semiconductor. Crucial to understanding this behavior is interrogating the interplay between solution processing parameters, the thin-film morphology, structural measurements, and device metrics within a rich process-structure-property space that can be time-consuming to measure. Unfortunately, capturing the exact location of behavior onsets, phase boundaries, and other transitions associated with improved performance can be challenging as they take place at very sensitive windows. The number of experiments required to fully understand the process-structure-property relationships quickly multiplies when subsequently exploring this composition-performance trend with respect to different polymers/solvents, or incorporating solution processing steps (e.g., UV-irradiation, aging, etc.). In this work, to aid in traversing the rich compositional landscape of polymer semiconductor/insulator blends, we incorporate a gradient film methodology through an automated flow system as a means of high-throughput experimentation. The screening of the P3HT/PS and DPP-DTT/PS composition libraries presented herein represents how the gradient film approach can facilitate preliminary exploration of process-structure-property relationships in polymer thin-film systems. The large amount of data generated from the gradient film libraries can then be integrated with data science techniques (e.g. machine learning) to model the relationship between polymer composition and OFET device performance. By leveraging the techniques reported herein, we envision the high throughput experimental screening approach to be broadly applicable to other solution-based thin-film systems, including organic device applications and beyond.

474e Decoding Optical Responses of Contact-Printed Arrays of Liquid Crystals Using Machine Learning: Detection of Aqueous Amphiphiles with Enhanced Sensitivity and Selectivity

Shiyi Qin, Fengrui Wang, Claribel Acevedo-Velez, Reid Van Lehn, David Lynn, Victor Zavala

Surfactants are amphiphilic molecules that contain hydrophilic head groups and hydrophobic tail groups [1]. Given this unique structure, surfactants and other amphiphilic molecules are used extensively in household products, industrial processes, and biological applications, and are also common environmental contaminants [2-5]; as such, methods that can detect, sense, or quantify them are of great practical relevance.
State-of-the-art methods for detecting natural and synthetic amphiphiles include mass spectrometry and high-performance liquid chromatography [4, 6]. While these methods provide excellent sensitivity and selectivity, they often require complex sample preparation procedures, costly laboratory infrastructure, and highly trained personnel.

To address these limitations, new approaches have been developed in recent studies, including many optical, electrochemical, and mechanical-based techniques [4, 7]. Among these new approaches, aqueous emulsions of thermotropic liquid crystals (LCs) [8-13] have gained growing attention because they can exhibit distinctive optical responses in the presence of surfactants. When coupled with machine learning techniques such as convolutional neural networks (CNNs), LC-in-water emulsions have emerged as sensitive, rapid, and inexpensive sensors or reporters of environmental amphiphiles [14-18]. However, many existing CNN-based LC sensing methods require the use of expensive instrumentation (e.g., flow cytometry [14]) or computational resources (e.g., videos [18]) for quantitative characterization, owing to variations in optical responses among individual LC droplets with different sizes in an LC-in-water emulsion.

Here, we report an LC-based surfactant sensing platform that takes a step toward addressing several of the abovementioned issues and can predict concentrations and types of model surfactants in aqueous solutions. Our approach uses microcontact-printed arrays of micrometer-scale droplets of thermotropic LCs and hierarchical CNNs to automatically extract and decode rich information about topological defects and color patterns available in optical micrographs of LC droplets to classify and quantify adsorbed surfactants. The microcontact-printed LC droplets are surface-immobilized and uniformly sized, reducing variations in optical responses. The proposed hierarchical CNN contains two levels of 2D CNNs that are trained on different data representations. The selection of the data representation for each level relies on the detection focus. The first level is intended to differentiate defect patterns, and therefore grayscale images, which highlight the contrast between light and shade and contain sufficient information for defect pattern classification. The second level, on the other hand, is forced to learn patterns that are less distinguishable and thus require color information (e.g., RGB). Overall, we show that the combination of microcontact-printed LC arrays and machine learning provides a convenient and robust platform that could prove useful for developing high-throughput sensors for on-site testing of environmentally or biologically relevant amphiphiles.

References

[1] Rosen, M. J., & Kunjappu, J. T. (2012). Surfactants and interfacial phenomena. John Wiley & Sons.
[2] Gallou, F., Isley, N. A., Ganic, A., Onken, U., & Parmentier, M. (2015). Surfactant technology applied toward an active pharmaceutical ingredient: more than a simple green chemistry advance.

Green Chemistry, 18(1), 14-19.
[3] Torchilin, V. P. (2001). Structure and design of polymeric surfactant-based drug delivery systems. Journal of controlled release, 73(2-3), 137-172.
[4] Kubicek-Sutherland, J. Z., Vu, D. M., Mendez, H. M., Jakhar, S., & Mukundan, H. (2017). Detection of lipid and amphiphilic biomarkers for disease diagnostics. Biosensors, 7(3), 25.
[5] Arora, J., Ranjan, A., Chauhan, A., Biswas, R., Rajput, V. D., Sushkova, S., ... & Jindal, T. (2022). Surfactant pollution, an emerging threat to ecosystem: Approaches for effective bacterial degradation. Journal of Applied Microbiology, 133(3), 1229-1244.
[6] Behrens, B., Helmer, P. O., Tiso, T., Blank, L. M., & Hayen, H. (2016). Rhamnolipid biosurfactant analysis using online turbulent flow chromatography-liquid chromatography-tandem mass spectrometry. Journal of Chromatography A, 1465, 90-97.
[7] Chen, Y. T., & Yang, J. T. (2015). Detection of an amphiphilic biosample in a paper microchannel based on length. Biomedical Microdevices, 17, 1-8.
[8] Lin, I. H., Miller, D. S., Bertics, P. J., Murphy, C. J., de Pablo, J. J., & Abbott, N. L. (2011). Endotoxin-induced structural transformations in liquid crystalline droplets. Science, 332(6035), 1297-1300.
[9] Carlton, R. J., Hunter, J. T., Miller, D. S., Abbasi, R., Mushenheim, P. C., Tan, L. N., & Abbott, N. L. (2013). Chemical and biological sensing using liquid crystals. Liquid crystals reviews, 1(1), 29-51.
[10] Agarwal, H., Nyffeler, K. E., Manna, U., Blackwell, H. E., & Lynn, D. M. (2021). Liquid Crystal-Infused Porous Polymer Surfaces: A "Slippery" Soft Material Platform for the Naked-Eye Detection and Discrimination of Amphiphilic Species. ACS applied materials & interfaces, 13(28), 33652-33663.
[11] Guo, X., Manna, U., Abbott, N. L., & Lynn, D. M. (2015). Covalent immobilization of caged liquid crystal microdroplets on surfaces. ACS Applied Materials & Interfaces, 7(48), 26892-26903.
[12] Manna, U., Zayas-Gonzalez, Y. M., Carlton, R. J., Caruso, F., Abbott, N. L., & Lynn, D. M. (2013). Liquid crystal chemical sensors that cells can wear. Angewandte Chemie International Edition, 52(52), 14011-14015.
[13] Ortiz, B. J., Boursier, M. E., Barrett, K. L., Manson, D. E., Amador-Noguez, D., Abbott, N. L., ... & Lynn, D. M. (2020). Liquid crystal emulsions that intercept and report on bacterial quorum sensing. ACS applied materials & interfaces, 12(26), 29056-29065.
[14] Jiang, S., Noh, J., Park, C., Smith, A. D., Abbott, N. L., & Zavala, V. M. (2021). Using machine learning and liquid crystal droplets to identify and quantify endotoxins from different bacterial species. Analyst, 146(4), 1224-1233.
[15] Cao, Y., Yu, H., Abbott, N. L., & Zavala, V. M. (2018). Machine learning algorithms for liquid crystal-based sensors. ACS sensors, 3(11), 2237-2245.
[16] Bao, N., Jiang, S., Smith, A., Schauer, J. J., Mavrikakis, M., Van Lehn, R. C., ... & Abbott, N. L. (2022). Sensing Gas Mixtures by Analyzing the Spatiotemporal Optical Responses of Liquid Crystals Using 3D Convolutional Neural Networks. ACS sensors, 7(9), 2545-2555.
[17] Smith, A. D., Abbott, N., & Zavala, V. M. (2020). Convolutional network analysis of optical micrographs for liquid crystal sensors. The Journal of Physical Chemistry C, 124(28), 15152-15161.
[18] Frazão, J., Palma, S. I., Costa, H. M., Alves, C., Roque, A. C., & Silveira, M. (2021). Optical gas sensing with liquid crystal droplets and convolutional neural networks. Sensors, 21(8), 2854.

474f Uni-MOF: A Universal Material Representationlearning Framework for Metal-Organic Frameworks

Jingqi Wang, Jiapeng Liu, Hongshuai Wang, Musen Zhou, Guolin Ke, Linfeng Zhang, Jianzhong Wu, Zhifeng Gao, Diannan Lu

Gas separation process is ubiquitous in industrial production, compared with conventional separation technologies, advanced nanoporous materials such as Metal-Organic Frameworks (MOFs) have advantages in terms of energy and separation efficiency due to their tunable pore structure and chemical affinities. The crucial issue is how to rapidly predict the gas adsorption capacity of MOF adsorbents, however, time-consuming synthesis and adsorption experiments as well as expensive simulations are not preferred. Therefore, we build a universal material learning framework based on transformer for MOFs (Uni-MOF). This pre-trained representation learning framework extracts structural information from hundreds of thousands of 3D materials with no labels, which enables precise prediction of gas adsorption amount based on small data sets and cross-system prediction scenarios. This pre-trained model is also versatile for screening MOF materials under different application criteria.

474g Shapelet-Based Defect Identification and Classification for Nanostructure Surface Imaging

Nasser M. Abukhdeir, Matthew P. Tino

Microscopy-based characterization techniques for nanomaterials are increasingly accurate and accessible to researchers. In order to determine structure/property relations, quantitative methods are needed to interpret microscopy images, and this is especially challenging for structured materials. For example, surface self-assembly processes produce microscopy images with complex spatially varying patterns that include large numbers of defects, both orientational and translational. Recent advancements in quantitative characterization of (nano)structured surfaces have enabled automated analysis of microscopy images. These methods include the use of local bond orientational order parameters and polar shapelets, focused on identification, classification, and quantification of local order. However, to date, there has been less focus on classification of defects in local order.

Defects in nanomaterials, including both translational (dislocations) and orientational (disclination) types, are topologically equivalent to those observed in crystalline phases. However, in soft materials the local structure of defects can vary substantially for equivalent topological charge. This is due to ordered phases in soft materials typically resulting from self-assembly of many constituent molecules, which allows for significant variation in local pattern length scale and the presence of local disorder within extreme pattern deformation.
Furthermore, the use of defects to manipulate and improve material performance has become increasingly recognized as an important aspect of material design, sometimes referred to as defect structure engineering. Therefore, the development of computational resources to both identify and classify these defects is imperative for quantification of local (nano)structure/property order.

In this work, a previously developed shapelet-based method for quantification of surface order is used to develop an unsupervised defect identification and classification method that is computationally more efficient than previous work. The previously developed shapelet-based method requires user input; specifically, the identification of a local pattern of interest within the target microscopy image. The presented method uses local pattern orientation with respect to a selected shapelet basis set in conjunction with a clustering scheme to automatically identify both translational and orientational defect structures without requiring information from the user; it also produces a defect identification output containing less noise and response variation compared to previous work. Identified defect structures are then classified based on shapelet response vectors, similar to the previously developed method, in order to classify them based on local pattern structure.

The defect identification and classification method is applied to a range of representative ordered surfaces resulting from self-assembly including stripe and hexagonal structures. Imaging from simulations and experimentation of surface self-assembly are analyzed using the presented method and compared to identification/classification conducted manually. The method is able to successfully identify and classify orientational and translational defects on a variety of structural patterns, depending on the resolution and noise present in the input microscopy image.

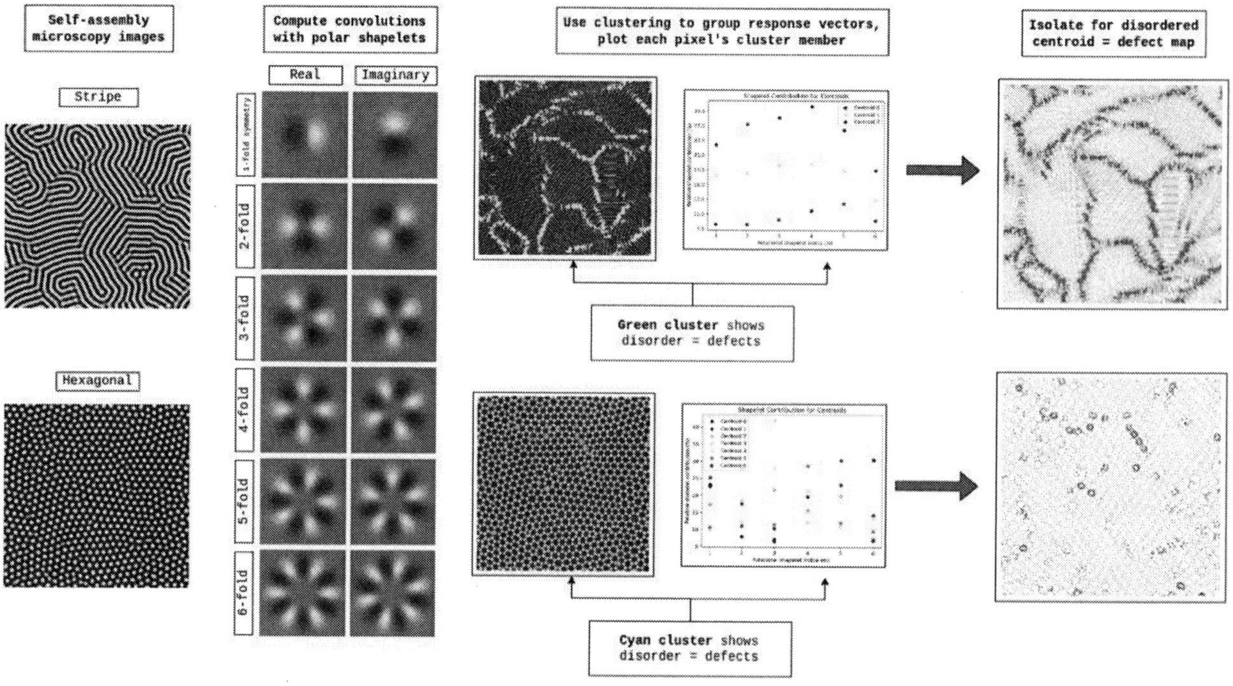

Self-assembly microscopy images
Stripe
Hexagonal
Compute convolutions with polar shapelets
Real
Imaginary
1-fold symmetry
2-fold
3-fold
4-fold
5-fold
6-fold
Use clustering to group response vectors, plot each pixel's cluster member
Green cluster shows disorder = defects
Cyan cluster shows disorder = defects
Isolate for disordered centroid = defect map

474h Predicting Pair Correlation Functions of Soft and Hard Materials Using Machine Learning

Kumar Ayush, Tarak Patra

The pair correlation function (PCF) is an important feature to study the structure of materials at the atomic or molecular level. It provides information about the probability of finding two particles at a certain distance from each other, and can be used to characterize the spatial arrangement of particles in a material. Soft materials, such as polymers, colloids, and biomolecules, are characterized by their flexibility and deformability. Their PCFs typically show long-range correlations and broad peaks, reflecting their disordered and heterogeneous nature. Hard materials, such as crystals, semiconductors, and metals, are characterized by their rigidity and strong interatomic interactions. Their PCFs typically show sharp peaks and short-range correlations, reflecting their ordered and homogeneous nature. Molecular simulations have been routinely used to estimate the PCF of materials. However, they are computationally very expensive, and estimating PCFs for the entire compositional space of a material remains a substantially challenging task. Here, we propose a machine learning pipeline (ML) for rapid prediction of PCF of a material based on the information of its composition. We conduct ~ 100 molecular simulations for a given material and calculate its PCFs for varying composition. We use these PCFs and their corresponding composition information to build, test and validate the ML method. Within this ML pipeline, the grayscale images of PCF of a material are encoded to a latent space using a convolutional neural network (CNN) autoencoder. Subsequently, a random forest regressor establishes a correlation between the composition of the material and the latent space representation of its PCF. This ML model is then used for predicting PCF for many unknown compositions of the material. We have successfully demonstrated the performance of our ML model for three representative cases – binary liquid mixtures, polymer nanocomposites and oxide glasses. This method is very generic and can accelerate the design, discovery, and fundamental understanding of composition-PCF relations of a wide range of soft and hard materials.

474i Exploring the Impact of Training Data Distributions on the Accuracy of Machine Learning Force Fields

Orlando A. Mendible Barreto

Neural network interatomic potentials (NNIPs) have become an increasingly popular method in molecular simulations because they provide accurate and computationally efficient evaluations of atomic interactions. In this study, we investigate the impact of the configurational sampling distribution on the accuracy of NNIPs for two molecular systems, butane and alanine dipeptide. Specifically, we examine scenarios where the range of the collective variables (CVs) required for free energy surface determination is not known a priori and investigate the effect of the training data distribution on the accuracy of NNIPs. To address this, we create representative datasets that mimic various distributions of configurations and use hyperparameter optimization to train the NNIP models while analyzing their dependence on system size and shape. Additionally, we propose a comprehensive testing procedure that evaluates energy/force predictions, molecular dynamics stability, structural properties, thermodynamic properties, and free energy surface determination using enhanced sampling techniques to validate the accuracy and stability of the models. Our study highlights the challenges associated with generating a training dataset with a sufficient representation of CVs for these molecular systems and provides a robust testing methodology for accurate free energy surface determination using NNIPs and enhanced sampling techniques when the required CV range is unknown.

474j Training Accurate and Physically Meaningful Machine Learning Force Fields for Water and Understanding Their Transferability

Tristan Maxson, Tibor Szilvasi

Water is an important material relevant to biology, chemistry, and catalysis among others, however, the accurate simulation of water is challenging because it requires long molecular dynamics simulations to equilibrate the system due to the complex interactions between the molecules. The MB-Pol potential has been developed to reproduce CCSD(T) quality results, but at a lower cost than full quantum mechanical simulations and allows to reproduce the ground truth for physical properties (e.g. density, compressibility, radial distribution function, etc.) which are typically inaccessible using quantum mechanical calculations due to long equilibration times (10 – 100 ns). Machine learning force fields (MLFFs) serve as an alternative to accelerate atomistic simulations by learning the complex set of interactions from existing data. We show that the Allegro extension of NequIP can accurately learn from liquid, gas, and solid phases of H_2O. The transferability of MLFF models trained on different phases is also explored to determine what information is crucial to the resulting MLFF. Importantly, we find that our MLFFs can describe physically meaningful many-body interactions that are not directly trained on and allow even faster simulations than MB-pol. In addition, MLFFs will be possible to combine with additional data to describe more complex systems than water only that are directly relevant to practical problems in biology, chemistry, and catalysis.

474k Learning Interatomic Forces from Experimental Measurements of Fluid Structure

Brennon L. Shanks, Harry W. Sullivan, Michael P. Hoepfner

The prediction of interatomic forces from experimental X-ray and neutron scattering data is a historic inverse problem in statistical mechanics. Accurate and robust reconstructions of interatomic forces could be used to create novel force fields for molecular simulations, study many-body effects in statistical ensembles, and advance our understanding of atomic structure and self-assembly of materials. However, there are no existing statistical theories or computational techniques that provide adequate inverse problem solutions for a wide range of physical systems. Furthermore, data-driven methods such as probabilistic machine learning are burdened by the high computational cost of evaluating models for atomic structures. In this study, we developed a novel Gaussian process surrogate model that vastly accelerates the estimation of structure data within a molecular simulation framework. We demonstrate that this method provides rapid and accurate evaluation of ensemble fluid structures that enables the application of probabilistic machine learning to optimize force fields from experimental scattering data with uncertainty quantification. These results suggest that machine learning accelerated interatomic force reconstruction from experimental diffraction data is now a viable tool for structural analysis and force field optimization.

474l Multi-Input E(n)-Graph Neural Networks for Learning Molecular Interaction Properties

Kieran Nehil-Puleo

Here we have developed a multi-input equivariant graph convolution-based deep learning model, named the MEGNN, for the prediction of chemical properties that result from the interaction of heterogeneous molecular structures. The model is an equivariant model which allows for the efficient learning of molecular properties by utilizing atomic spatial coordinates to make predictions that are agnostic to translations, rotations, and reflections in input molecules. We applied this model to predict the frictional properties between polymer brush monolayers with different functional end-groups attached to the ends of the brushes and achieved improvements to the prediction of these tribological properties. The MEGNN enables conditioned property prediction, meaning properties can be predicted qualified on their interaction with another molecule. In this talk I will discuss some of the technical details of this model and their relevance to chemistry.

420a Predicting Phase-Specific Polymerization Thermodynamics for Circular Polymer Design Using Machine Learning

Shivani Kozarekar, Linda Broadbelt

The swift rise in plastics production and usage, alongside inadequate recycling rates, has created a demand for sustainable materials that are easy to recycle. Intrinsically circular polymers (ICPs) are a promising class of materials that can be recycled at mild conditions through chemical conversion to their pure monomer form, making them an attractive alternative to conventional plastics. Designing ICPs computationally requires precise knowledge of the polymerizing system's thermodynamics, which are difficult to predict using molecular modeling methods such as density functional theory (DFT). This complexity arises from the interplay between monomer structure, polymer chain length and phase, with the entropic contributions particularly challenging to capture accurately. To rein in the high-dimensional nature of this problem, we propose using machine learning (ML) methods to predict thermodynamics of polymerization using features that describe system-wide interactions while staying within the constraints of the design space, focusing initially on the enthalpy of polymerization. This approach uses an experimental data set collected from literature and insights gained from simulation to select a relevant feature set. The models are benchmarked using a variety of ML workflows. The overall workflow for this approach is outlined in Figure 1. Our model can predict the enthalpy of polymerization for a given monomer structure and phase/solvent combination. By combining this machine learning approach with ML-informed chemical network expansion methods, we aim to design novel ICPs with desirable thermodynamic properties for experimental synthesis, with the goal of contributing to plastic waste reduction.

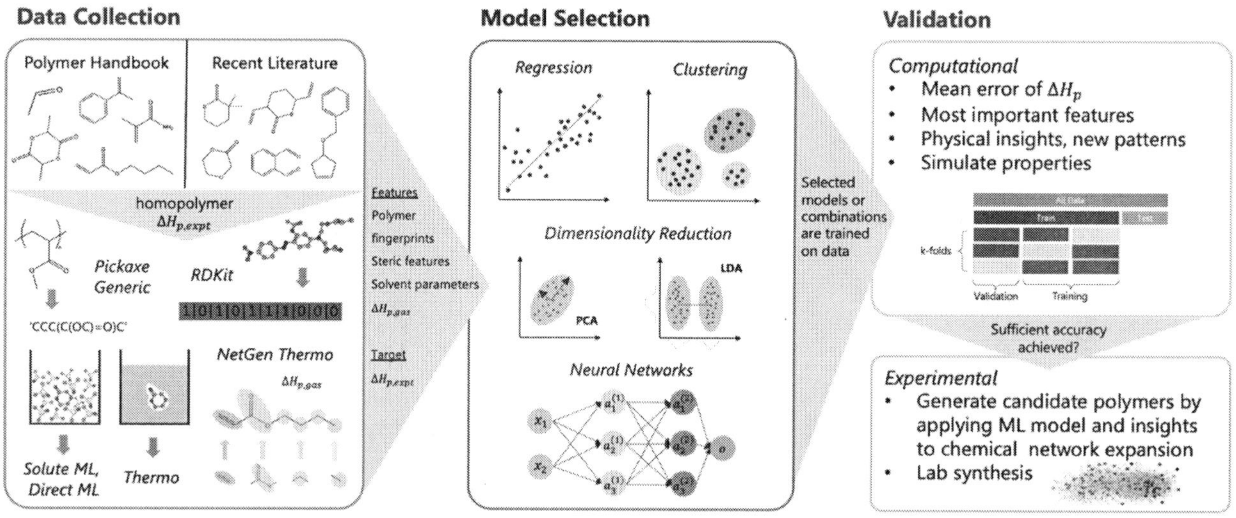

Figure 1. This schematic represents the machine learning workflow for predicting thermodynamic properties of ICPs.

420c An Autonomous Microreactor Platform for the Optimisation of Kinetic Parameters for Fast Liquid/Liquid Reactions

Anna Katsarou, Arun Pankajakshan, Jonathan Lefebvre, Julia Hofinger, Federico Galvanin, Alexei A. Lapkin

One of the most complex topics for design of multiphase liquid/liquid processes is the development of reliable and accurate kinetic models, since reaction kinetics and mass transfer are both controlling the process.[1] In order to develop intrinsic kinetic models, a researcher needs to ensure that the reaction rate is not limited by mass transfer phenomena, identify the model structure that can best describe the reaction chemistry and obtain precise kinetic parameters.

Batch processing is commonly used to obtain kinetic models. However, traditional stirred-tank reactors can be adversely affected by inefficient heat and mass transfer, safety concerns in case of exothermic reactions and long processing times overall because each batch needs to be processed separately in order to separate and analyse the products. Benefiting from their smaller reactor volumes, continuous flow microreactors can alleviate these challenges. The higher surface area-to-volume ratios allow for more efficient mass and heat transfer, safer usage of hazardous chemicals and precise control over reaction conditions.[2]

However, the effectiveness of model discrimination (i.e. the identification of the appropriate set of equations defining the model) between candidate models and the precise estimation of kinetic parameters depends on the conduction of properly designed experiments. Otherwise, less informative experiments may result in a waste of resources and time. Model-based design of experiments (MBDoE) is a tool for efficient development of deterministic models, which relies on *prior* knowledge of the model structure of the underlying system and initial parameter estimates to design an optimal experiment.[3] The further integration of MBDoE algorithms with automated flow platforms and inline analytical techniques is a powerful tool for obtaining intrinsic reaction kinetics in a very efficient way.[4] In such close-loop experimental systems, the MBDoE algorithm designs an experiment, which is executed automatically, and data collected in real time from the analytical instrument are interpreted by the algorithm to design the next experiment. This takes place iteratively until the optimization criteria are satisfied. This process aims to improve the way we obtain robust kinetic models by reducing experimentation time, overall cost and enhancing process sustainability.

This work shows the workflow for obtaining reliable and accurate kinetic models for fast liquid/liquid reactions by developing a close-loop experimental system. A reactive chemical system of interest is the heterogeneous nitration of 2-nitrotoluene (2-MNT), since it is a very fast reaction,[5] highly exothermic,[6] and reaction products, 2,4-dinitrotoluene (2,4-DNT) and 2,6-dinitrotoluene (2,6-DNT), are important intermediates in fine chemical industries.[7] In this work, we utilize an autonomous flow platform equipped with continuous flow micromixers to achieve efficient mixing and ensure process safety. Work-up steps are performed in-line and phases are separated based on membrane separation. Organic phase passes through a benchtop NMR for quantification of reaction products and data are processed by an MBDoE algorithm to design the next experiment with the objective of achieving a precise estimation of kinetic parameters for a pre-selected kinetic expression.[8]

References: [1] *React. Chem. Eng.*, 2019, 4, 235, [2] *Iscience*, 2022, 25, 103892, [3] *Chem. Eng. Sc.*, 2008, 63, 4846 – 4872, [4] ***React. Chem. Eng.***, 2019,**4**, 1623-1636, [5] *React. Chem. Eng.*, 2022, 7, 111-122, [6] *Beilstein J. Org. Chem.*, 2014, 10, 405-24, [7] *Phys. Chem. Chem. Phys.*, 2021, 23, 4658–4668, [8] *Chem. Eng. J.*, 2019, 377, 120346

420d Automatic Interpretation and Control of X-Ray Diffraction Experiments

Chris Bartel, Nathan Szymanski, Yan Zeng, Mouhamad Diallo, Qingsong Tu, Haegyeom Kim, Gerbrand Ceder

The emergence of high-throughput quantum chemical calculations has accelerated the rate at which we can predict new materials for various applications (batteries, solar cells, catalysts, etc.), but the successful synthesis of these materials has often become the slow step in materials design. Autonomous laboratories hold the potential to systematically explore various synthesis routes to new materials, alleviating the painstaking manual trial-and-error approach. However, for an autonomous laboratory to work for inorganic synthesis, we need a method to assess the success of a given synthesis effort without any human intervention. Powder X-ray diffraction (XRD) is the workhorse technique for determining the outcome of materials synthesis. In this talk, I will show how convolutional neural networks can be trained to automatically interpret XRD patterns and identify the phases present in realistic mixtures of crystalline solids. I will also discuss how these same models can be leveraged to adaptively control the XRD experiment itself, improving the quality of predictions and enabling the detection of short-lived intermediates that form during synthesis.

420e Active Learning for Data-Efficient Training of Machine Learning Models to Predict Adsorption in Metal-Organic Frameworks (MOFs).

Etinosa Osaro, Fernando Fajardo-Rojas, Diego Gomez Gualdron, Yamil Colón

Metal-organic frameworks (MOFs) are a promising class of porous, crystalline materials for numerous applications. For instance, a MOF with the "right" adsorption properties could enable replacing a given thermal-based, chemical separation process with and adsorption-based one, which could in turn bring up a 10-fold increase in energy efficiency. As chemical separation account for roughly 15% of U.S. energy usage, and about 80% of these separations are done thermally, finding the "right" MOF for each separation could potentially reduce U.S. energy expenditure by around 11%. Given *i)* the overwhelmingly large MOF "design space" (with trillions of potential designs), and *ii)* the thousands of chemical separations, each which could be potentially performed at a variety of different operating conditions (OC, e.g., temperature, pressure, relative proportion of components), one can imagine that computation is to play a central role in identifying for each chemical separation the most promising MOF with the corresponding "optimal" operating condition.

The challenge is that classical simulations methods to predict adsorption (e.g., grand canonical Monte Carlo (GCMC)) are just "fast enough" to make thousands to hundreds of thousand adsorption predictions in a reasonable timeframe. However, finding the optimal MOF-OC combination for each chemical separation of interest is a task that would probably entail trillions of adsorption predictions. Thus, faster methods such as machine learning (ML) are better poised to take such task. In earlier work, some of us demonstrated the ability of multilayer perceptron (MLP) models to learn to predict adsorption at multiple conditions, for multiple molecules when provided with GCMC-generated training data for adsorption of different molecules at different pressures in different MOFs. However, this demonstration was limited to near-spherical, non-polar molecules, and extension to a wider class of molecules requires increasing the diversity and size of the training data. However, due to the computational resources needed to generate training data and training the ML model, there is a critical need to keep training dataset as small as possible.

Active learning (AL) can play a very important role in efficiently and "smartly" navigate the "adsorption space" to limit the burden on data generation while enabling the training of highly predictive ML models. In this work, we first establish the implementation of a Gaussian process regression (GPR) framework to model pure component adsorption of nitrogen at 77K from 10^{-5} to 1 bar, methane at 298K from 10^{-5} to 100 bar, carbon dioxide at 298K from 10^{-5} to 100 bar, and hydrogen at 77K from 10^{-5} to 100 bar on eleven diverse sets of MOFs. In this GPR framework, a first model is trained with an initial data set known as the "prior." Then subsequent models are retrained upon subsequent addition of adsorption data to the dataset, which is decided by the uncertainty of the GP model evaluated on a new data set. Here, we tested three different "prior" selection schemes and make a recommendation on the best prior selection scheme for 44 adsorbate-adsorbent pairs. Recommendation is primarily based on the mean absolute error and the total amount of data points required for convergence of the prediction of the ML model.

Upon establishing the GPR framework, we demonstrated the application of the methodology to include alchemical molecules. These hypothetical species can be characterized by two main features: intermolecular potential parameters (e.g., well-depth and the distance at which the intermolecular potential between two particles is zero); and intra-molecular properties, such as bond length and charges.

A previously developed MLP model trained on GCMC data points (approximately 5 million) obtained from 1800 topologically and chemically diverse ToBaCCo generated MOFs using several single- and multiple-site alchemical species at different fugacities has led to the progress in adsorption studies providing accurate results for a diverse set of real molecules. Using the established AL framework and the developed MLP model as a substitute for GCMC, we show we can make accurate GPR models that predict the isotherm of all alchemical species across these 1800 diverse MOFs using a different set of test-data set including the fugacity and alchemical parameters. Our results show we saved 57.5% of the data, indicating that only around 2.2 million simulations are needed to train a new MLP model for adsorption.

420f Molecular Weight Distribution Modulation and Mechanical Property Prediction of Linear Polymers: Combining Molecule Simulation and Machine Learning

Yue Fang, Hanyu Gao

The molecular weight distribution (MWD) of linear polymers plays a critical role in determining their mechanical properties, which are essential for designing materials that exhibit high tensile strength and excellent rheological property and processability[1]. However, conventional experimental methods for studying the specific effects of MWD on mechanical properties can be time-consuming and labor-intensive. Inspired by structure–property framework with organosilica synthesis,[2] we propose an integrated design that combines kinetic Monte Carlo (KMC), molecular dynamics (MD) simulation, and machine learning (ML) to efficiently modulate MWD and uncover the intricate relationship between MWD and the resulting mechanical properties. First, we employ KMC simulations to generate various linear polymer with different MWD by adding initiators and monomers in a semi-batch manner. The simulated KMC results are validated by comparing them to published experimental data.[3] Next, the MD simulations are performed based on the KMC results to investigate the effects of different structural parameters on the mechanical properties, such as interaction, bond energy, average chain length, and polydispersity index (PDI). The results show that strong interactions and bond energies enhance the Young's modulus of the system due to more significant bond orientation and chain entanglement. The impact of chain length and PDI on mechanical properties are also investigated, and the results show that their effects on mechanical properties are non-monotonous. Furthermore, we build machine learning models based on molecular dynamics simulation results to link the structural parameters to the mechanical properties. These models predict the mechanical properties of the polymer, such as tensile strength, Young's modulus, and elongation at break. Our results demonstrate that the proposed machine learning models exhibit high predictive accuracy and robustness. This study showcases the potential of integrating molecular simulations and machine learning techniques for efficiently modulating MWD and predicting the mechanical properties of linear polymers. The proposed method can significantly expedite the development of new polymeric materials with customized mechanical properties, paving the way for their use in a diverse range of fine and everyday chemical applications.

References:

(1) Gentekos, D. T.; Sifri, R. J.; Fors, B. P. Controlling Polymer Properties through the Shape of the Molecular-Weight Distribution. *Nat Rev Mater* **2019**, *4* (12), 761–774.

(2) De Keer, L.; Kilic, K. I.; Van Steenberge, P. H. M.; Daelemans, L.; Kodura, D.; Frisch, H.; De Clerck, K.; Reyniers, M.-F.; Barner-Kowollik, C.; Dauskardt, R. H.; D'hooge, D. R. Computational Prediction of the Molecular Configuration of Three-Dimensional Network Polymers. *Nat. Mater.* **2021**, *20* (10), 1422–1430.

(3) Gentekos, D. T.; Dupuis, L. N.; Fors, B. P. Beyond Dispersity: Deterministic Control of Polymer Molecular Weight Distribution. *J. Am. Chem. Soc.* **2016**, *138* (6), 1848–1851.

420g Active Learning Enhanced Screening of CO_2 Capture Materials

Anantha Sundaram, Yogesh V. Joshi, Ashish B. Mhadeshwar, Jayashree Kalyanaraman, Joseph Falkowski, Jonathan Szlachta, Rodrigo Neumann, Binquan Luan, Ricardo Luis Ohta, Felipe Lopes Oliveira, Mathias Steiner

In this work, we have developed and demonstrated an end-to-end modeling framework for screening of solid sorbent materials for CO_2 capture. This framework is a wholistic consideration of all the scales from which the performance of a sorbent material arises, starting from molecular dynamics, adsorption, and diffusion phenomena to process related interactions such as gas composition and contamination. Each of these components is tied to structural motifs of the sorbent material with the aid of machine learning models in an active learning process. Multiple elements of unsupervised and supervised learning are folded into this approach where the clusters of data, aggregates of features, model structure and parameters are continuously learnt starting from a small set of diverse adsorbent structures. Molecular simulations and process models are directed to evaluate better performing, synthetically promising structures by a multi-objective screening procedure that also weighs in the variable uncertainty in machine learning (ML) model predictions across different structures. This stratified approach allows us to accelerate identification of promising candidate sorbents that satisfy multi-scale end-to-end criteria starting from a very limited dataset. The accuracy and confidence of the ML-models are also improved through each cycle. We believe this approach of incrementally blending in knowledge through an integrated active learning cycle can be applied successfully to a wide range of material discovery problems.

420h Advances Featurization and Characterization of MOF Pores for Adsorption Applications

Arun Gopalan, Kaihang Shi, Lev Sarkisov, Randall Snurr

Short summary: The cavities of a crystalline nanoporous material can be represented efficiently as a network of interconnected pockets called a 'pore graph' by using a combination of 3D image processing and some novel algorithms. The core workflow itself is quite potent for solving characterization problems such as window detection and pocket types detection in materials. Further, pore graphs can be flexibly annotated to make the most out of existing cavity descriptors for supervised and unsupervised learning of adsorption properties.

Modern day computational materials design

State-of-the-art approaches to computational discovery of MOFs can be thought of in terms of three key components, as shown in Figure 1

- *A representation engine*: A way to represent MOFs E.g., Network of atoms and bonds, a connected string of building blocks etc.
- *A prediction engine*: A method to predict the desired property E.g., Molecular simulations, ab initio calculations, machine learning etc.
- *A structure sampler/generator*: A plan to search the phase space of all possible MOFs E.g., Databases like CoRE-MOFs, ToBaCCo etc. or, an optimizers based on Genetic Algorithm, Bayes etc.

Main challenge in representing MOFs for adsorption

How we represent materials is key to the efficiency of all three steps above. In adsorption, the properties are more strongly correlated to the energy and connectivity of the voids in the material. On one hand, the popular string and graph representations of MOFs (e.g., structure graph, MOFId etc.) primarily gear towards the chemical structure of the MOF and are far too detached to describe the 3D cavities sufficiently well. On the other hand, 3D grid (image) representations of materials are very comprehensive but are too computation and memory intensive to be compiled into a database to be used in high throughput. Hence, we present an intuitive and robust workflow for representing the 3D confining environment of a crystalline nanoporous materials efficiently. The result is a '*pore graph*', which is flexible enough to retain any cavity information relevant to adsorption, while being easy to compute, store and retrieve when used in characterization or machine learning problems.

The pore graph workflow

The void space in a material is converted into a network of interconnected pockets, using 3D image segmentation. As an example, let's look at this workflow applied to Cu-BTC (HKUST-1) as shown in Figure 2. First, the 3D distance map (Figure 2, Step 1) of the material is generated from its CIF file (Figure 2, Step 0), where each voxel is assigned a distance value to the nearest atom surface. The distance map is then thresholded to generate a binary image, where the void space is represented by 1s and the solid material by 0s (not shown in Figure 2). The binary image is then segmented into regions about the local distance maxima using Watershed segmentation (Figure 2, Step 2), which is then converted into a network of interconnected pockets, called the *pore graph* (Figure 2, Step 4). Each node on this graph represents a pocket in the material and each edge represents the connectivity between two pockets through one or more pore windows. In crystalline materials, it is necessary to account for PBC correctly while describing cavities, to average the properties efficiently. However, given the variety of ways in which the unit cell can cut through pockets, this is quite non-trivial. Hence, we implemented novel algorithms to account for PBC correctly. The result is a regrouped list of region indices called periodic groups, that tells which regions/pockets should be put together to form a complete pocket (Figure 2, Step 3). For example, in the case of Cu-BTC, all the 8 regions at the corners of the unit cell (Figure 2, Step 3) should be combined to get a complete pocket, hence its periodic group has 8 entries. Instead, if a pocket is fully inside the unit cell, then it has only one entry in its periodic group. This information is then annotated on to the nodes of the pore graph.

Two proof-of-concept characterization applications

Breaking down the void space into constituent pockets serves as a template to get the most out of simple pore descriptors and describe the cavities better. As examples, solutions to two challenging characterization problems using this workflow are described below:

1. *Detect pore windows in materials:*

When two regions are connected, the outer boundary of one region, overlaps with the inner boundary of the other region. The windows can be detected as the centroid of this shared outer boundary. If two regions are connected through multiple windows, an agglomerative clustering step is used to group the shared boundary into separate windows. For example, Figure 3, Application 1 shows the windows in Cu-BTC detected and overlayed on its distance isosurface and its pore graph.

2. *Detect the types of pockets in a material:*
   Once PBC is applied correctly, a simple point-to-centroid distance histogram is calculated for each periodic group. We then used this simple, geometric descriptor to detect the different types of pockets in the material using agglomerative clustering. For example, Figure 3, Application 2 shows the three types of pockets detected in Cu-BTC and their corresponding size histograms .

Concluding remarks

Many other characterization problems can also be solved using this workflow including the breakdown of total simulated isotherms into regional isotherms using adsorption snapshots and comparing shapes of pockets across different materials and/or molecules using persistent homology. Apart from pure characterization, the pore graphs can be annotated with any node (pocket) or edge (connection) attribute (LCD, window size, energy histogram etc.) and fed into graph ML algorithms to predict complex adsorption properties that are often inaccessible to simpler architectures. This is much more efficient than training CNNs with cumbersome 3D images of materials. In summary, pore graphs help us understand the cavities in a material better by representing them as a network of interconnected pockets. This reveals the underlying similarities and opens the possibility for solving otherwise difficult characterization problems, and adsorption property prediction via machine learning.

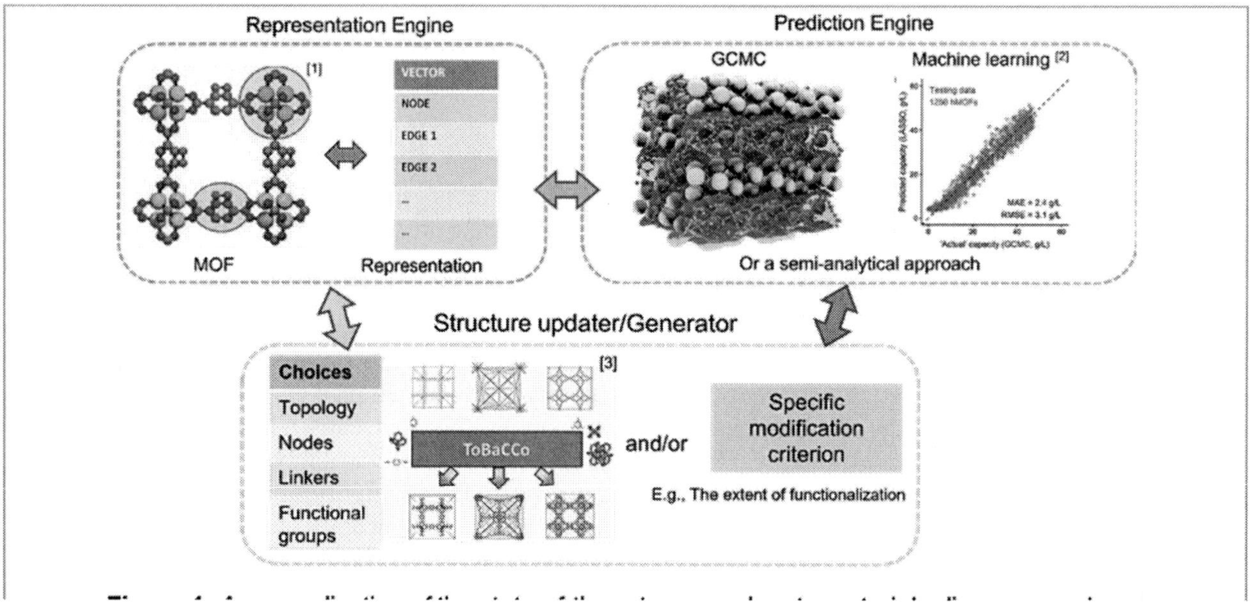

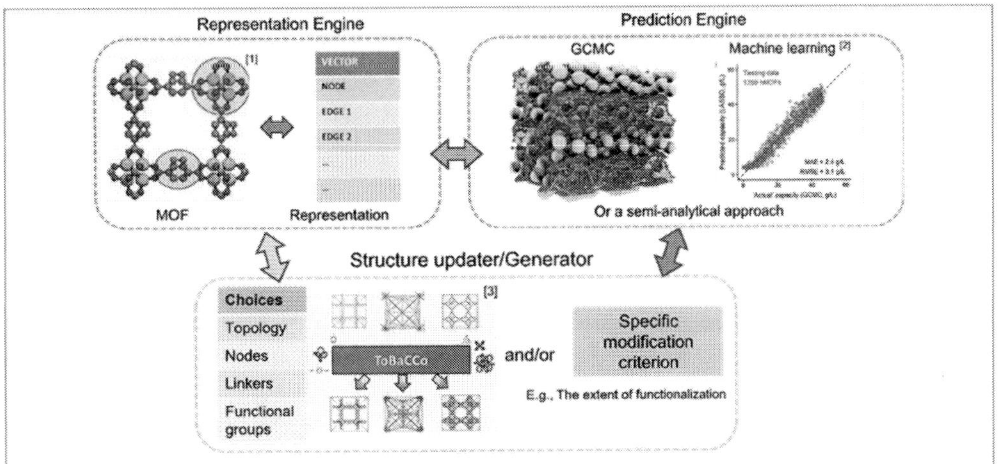

Figure 1: A generalization of the state-of-the-art approaches to materials discovery, using some popular choices as the three components, Representation, Prediction and Generation

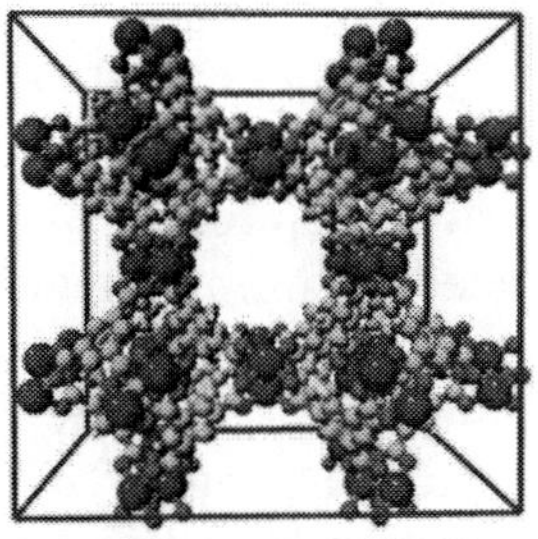

Step 0: Start from the CIF file of the material

Step 1: Compute the distance grid and detect the local maxima. 2 Å isosurface is shown here. Points of different colors represent local maxima of different values (each maxima corresponds to distinct pocket)

Step 2: Apply 3D Watershed region segmentation to break down the void space into distinct pockets about the detected local maxima.

Core workflow

Figure 2: Steps in the workflow used in generating graph representation of the cavities in a material using CuBTC as an example starting from the CIF file.

- Distance grid
- Region segmentation
- Accounting for PBC
- Constructing and annotating pore graphs

**steps are detailed in the main text*

Step 3: Account for PBC. Group pockets such that all regions in a group together forms a complete pocket. Note that all the regions that form a periodic group are now of the same color.

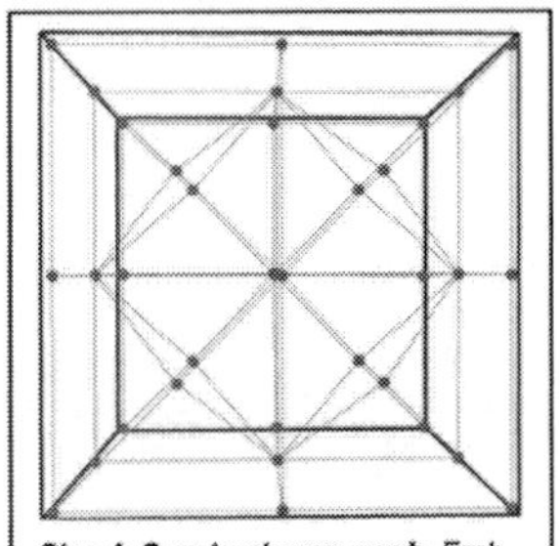

Step 4: Construct pore graph. Each node in the graph corresponds to the pocket, and connectivity of the pockets is denoted by edges. Pore graph is annotated with useful information about the cavities, including PBC information.

Application 1: *Detect windows between connected pockets*

Windows are located (shown in red) as the centroid(s) of the shared boundary between two connected regions, and overlayed on the isosurface and pore graph.

Application 2: *Detect pocket types using Agglomerative clustering*

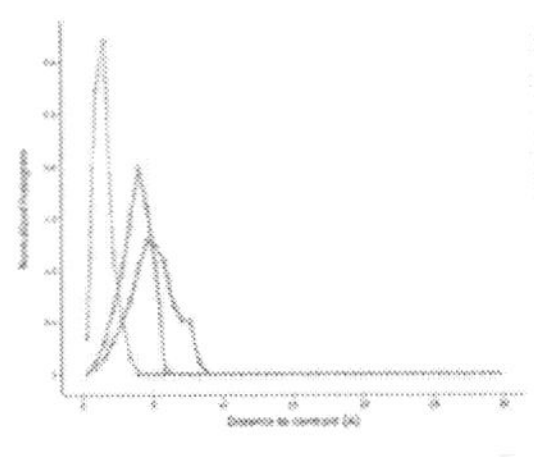

Simple cavity size histograms used in grouping the pockets into types. Color indicates the type of pocket, consistent with figure on the right.

Pockets are colored by the pocket type detected. There are three distinct types of pockets in Cu-BTC, shown in gray (r=5.5 Å), turquoise (r=5.1 Å) and light blue (r=2.66 Å).

Figure 3: Two challenging characterization problems using this workflow. *Left:* detecting pore windows in the material(shown as red dots) *Right:* Detect the different types (geometric) of pockets in a material via agglomerative clustering of simple cavity histogram features

420i Learning Material Stability Scores from Computationally Generated Kinetic Data

Veerupaksh Singla, Qiyuan Zhao, Brett Savoie

Despite the rapid progress in computational property prediction of novel materials, the absence of computational methods to predict kinetic degradation characteristics represents a costly bottleneck to inserting new materials into applications. While there are experimental stress testing methods to characterize material stability, such methods sacrifice the time and cost benefits of computational design and only supply information retrospectively rather than during the materials design phase. Additionally, on-the-fly screening of prospective materials is critical to exploratory material development workflows. Thus, it is crucial to have inexpensive computational tools capable of characterizing chemical degradation kinetics in advance of costly synthesis and stress testing to provide qualitative guidance on the susceptibility of a material to various stressors. Here, we report the development of an inexpensive scalar surrogate called the ***Stability Score***, which reflects the stressor-specific stability rank order of materials. We use the thermal degradation of alkanes as an exemplar system to set up the methodology and show how we tackled the problem of kinetic data scarcity. Using a model reaction-based approach for reaction condensation and a depth-wise kinetically pruned reaction generation approach for network reduction, thermal degradation pathways with ~1B potential reactions were reduced to computationally tractable 59 DFT level activation barrier calculations and systems of ~1k kinetically relevant reactions. We generated thermal degradation half-lives for 32k alkanes, benchmarked the stability trends with the limited available experimental kinetic data, and derived non-trivial insights from experimental thermodynamic data. Using this half-life data for training and testing, we trained a multi-layer perceptron (MLP) with a modified hinge-loss function to predict a scaled rank order of these half-lives ***(Stability Score)***, and compared the performance with a more complex Message Passing Neural Network (MPNN). These models were successful on extrapolation tasks on multiple train-test data splits. We also demonstrate a simple material design problem that uses Stability Scores to guide the search. This work provides first-of-its-kind evidence that kinetic data can be generated in a robust manner to train deep learning models to learn the relative kinetic information between different molecules to predict chemical stressor-specific stability and even extrapolate that information to aid material design. Furthermore, this work sets the initial groundwork for exploring the degradation of more complex chemistries in the future.

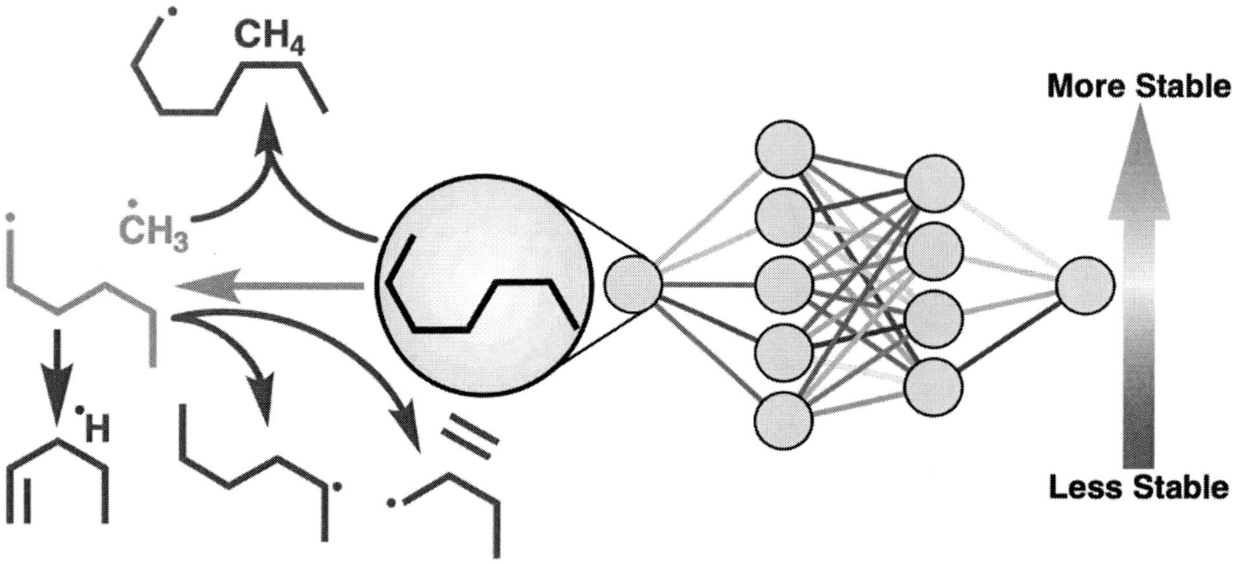

420j Mispr: A High-Throughput Multi-Scale Infrastructure for Automating Molecular Simulations

Nav Nidhi Rajput, Rasha Atwi, Matthew Bliss

In this talk, I will present a robust open-source high-throughput multi-scale computational infrastructure coined MISPR (Materials Informatics for Structure-Property Relationships) developed by our lab that seamlessly integrates classical molecular dynamics (MD) simulations with density functional theory (DFT).[1] By enabling high-performance data analytics and coupling between different methods and scales, MISPR addresses critical challenges arising from the needs of automated workflow management and data provenance recording. The major features of MISPR include automated DFT and MD simulations, error handling, derivation of molecular and ensemble properties, and creation of output databases that organize results from individual calculations to enable reproducibility and transparency. I will describe fully automated DFT and MD workflows implemented in MISPR to compute various electronic properties such as nuclear magnetic resonance chemical shift, binding energy, bond dissociation energy, and redox potential with support for multiple methods such as electron transfer and proton-coupled electron transfer reactions.[2] The infrastructure also enables the characterization of large-scale ensemble properties by providing MD workflows that calculate a wide range of structural and dynamical properties in liquid solutions and at solid-liquid interfaces. At the backend, the infrastructure interfaces with the Gaussian [3] software which enables electronic structure calculations of chemical systems, and LAMMPS [4] open-source code for MD simulations. In addition to its ability to handle the generation of input/data files and parsing of output files from the mentioned computational software, the infrastructure allows management of the collected data and storing it in MongoDB [5], a NoSQL database program using JSON-like documents with flexible schema. Each derived property is saved in its own collection with auxiliary information like molecular metadata (smiles representation, chemical formula, ...), which makes it easy to query and data-mine structure-property relationships. The user can tune the calculations by overriding default workflow parameters, for example, the functional and basis set, by-passing selected steps, or packing many jobs over multiple nodes for supercomputing resources. LAMMPS workflows allow the execution of MD simulations in different ensembles and analysis of the dumped trajectories for various dynamical and structural properties using another open-source codebase developed by our lab called **MDPropTools.**[6] **MDPropTools** is a powerful standalone in-house suite of Python-based post-processing routines, to perform statistical analysis of thermodynamic, structural, and dynamical properties of MD simulations of liquids and solid-liquid interfaces.[6] Using MISPR and MDPropTools, we published the first publicly available database, **ComBat** ~2,000 computed QC and MD properties for reported LSB electrolytes composed of solvents spanning 16 different chemical classes.[7, 8] The data generated from high-throughput physical models are then used to train **machine learning (ML) models** that will allow exploration of a larger chemical and parameter space at a significantly higher speed to establish structure-property relationships in multicomponent solutions. MISPR employs the methodologies of materials informatics to facilitate understanding and prediction of phenomenological structure-property relationships, which are crucial to designing novel optimal materials for numerous scientific applications and engineering technologies.

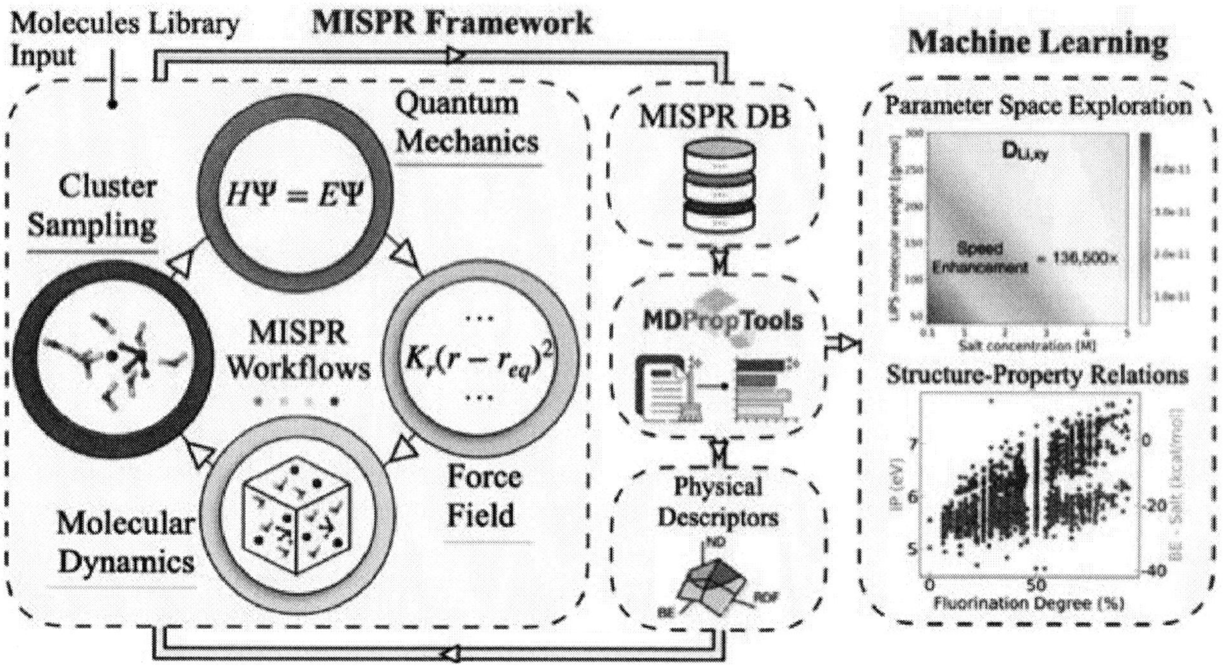
Molecules Library Input
MISPR Framework
Machine Learning
Quantum Mechanics
Cluster Sampling
$H\Psi = E\Psi$
MISPR Workflows
$K_r(r - r_{eq})^2$
Force Field
Molecular Dynamics
MISPR DB
MDPropTools
Physical Descriptors
Parameter Space Exploration
Speed Enhancement = 136,500×
Salt concentration (M)
Structure-Property Relations
IP (eV)
BE - Salt (kcal/mol)
Fluorination Degree (%)

AUTHOR INDEX